Natural Computing Series

Series Editors: G. Rozenberg
Th. Bäck A.E. Eiben J.N. Kok H.P. Spaink

Leiden Center for Natural Computing

For further volumes:
www.springer.com/series/4190

Frank Neumann · Carsten Witt

Bioinspired Computation
in Combinatorial Optimization

Algorithms and Their
Computational Complexity

 Springer

Dr. Frank Neumann
Max Planck Institute for Informatics
Dept. of Algorithms and Complexity
Campus E14
66123 Saarbrücken
Germany
frank.neumann@mpi-inf.mpg.de

Dr. Carsten Witt
Technical University of Denmark
Dept. of Informatics
and Mathematical Modelling
Richard Petersens Plads
2800 Kgs. Lyngby
Denmark
cawi@imm.dtu.dk

Series Editors

G. Rozenberg (Managing Editor)
rozenber@liacs.nl

Th. Bäck, J.N. Kok, H.P. Spaink
Leiden Center for Natural Computing
Leiden University
Niels Bohrweg 1
2333 CA Leiden, The Netherlands

A.E. Eiben
Vrije Universiteit Amsterdam
The Netherlands

ISSN 1619-7127
ISBN 978-3-642-16543-6   e-ISBN 978-3-642-16544-3
DOI 10.1007/978-3-642-16544-3
Springer Heidelberg Dordrecht London New York

ACM Computing Classification (1998): F.2, G.1, G.2, I.2

Cover design: KünkelLopka GmbH, Heidelberg

Printed on acid-free paper

Springer is part of Springer Science+Business Media (www.springer.com)

To Aneta, Linda, Michelle and Ying

Foreword

Bioinspired computing is successful in practice. Over the past decade a body of theory for bioinspired computing has been developed. The authors have contributed significantly to this body and give a highly readable account of it. (*Kurt Mehlhorn, Max Planck Institute for Informatics, and Saarland University, Germany*)

Bioinspired algorithms belong to the most powerful methods used to tackle real world optimization problems. This book gives such algorithms a solid foundation. It presents some of the most exciting results that have been obtained in bioinspired computing in the last decade. (*Zbigniew Michalewicz, University of Adelaide, Australia*)

This book presents a most welcome theoretical computer science approach and perspective to the design and analysis of discrete evolutionary algorithms. It describes the design and derivation of evolutionary algorithms which have precise computation complexity bounds for combinatorial optimization. The book should appeal to researchers and practitioners of evolutionary algorithms and computation who want to learn the state of the art in evolutionary algorithm theory. (*Una-May O'Reilly, CSAIL, MIT, USA*)

The evolutionary computation community has been in need of rigorous results concerning the computational complexity of their approaches for decades. This is the first textbook covering such a fundamental topic. It provides an excellent overview of the state of the art in this research area, in terms of both the results obtained and the analytical methods. It is an indispensable book for everyone who is interested in the foundations of evolutionary computation. (*Xin Yao, University of Birmingham, UK*)

Preface

Inspiration from biology has led to several successful algorithmic approaches. Such methods are frequently used to tackle hard and complex optimization problems. Biologically inspired algorithms such as evolutionary algorithms and ant colony optimization have found numerous applications for solving problems from computational biology, engineering, logistics, and telecommunications. Many problems arising in these application domains belong to the field of combinatorial optimization. Bio-inspired algorithms have achieved tremendous success when applied to such problems in recent years.

In contrast to many successful applications of bio-inspired algorithms, the theoretical foundation of these algorithms lags far behind their practical success. This is mainly due to the fact that these algorithms make use of random decisions in different modules. This leads to stochastic processes that are hard to analyze. This book treats bio-inspired computing methods as stochastic algorithms and presents rigorous results on their runtime behavior.

The book is meant to give researchers a state-of-the-art presentation on theoretical results of bio-inspired computing methods in the context of combinatorial optimization. Furthermore, it can be used as basic material for courses on bio-inspired computing that are meant for graduate students and advanced undergraduates.

The book is organized into three parts. It starts with a general introduction into bio-inspired algorithms and their computational complexity. Later on, different methods that have been developed in recent years are presented in a comprehensive manner. Afterwards, we present some of the major results that have been achieved in the field of single-objective optimization. We consider different problems such as minimum spanning trees, maximum matchings, and the computation of shortest paths. After these studies, we turn to multi-objective optimization. We tackle classical multi-objective problems such as the computation of multi-objective minimum spanning trees as well as show that multi-objective approaches lead provably to better algorithms for classical single-objective problems.

Taking the book as basic material for a course on theoretical aspects of bio-inspired computing, we suggest you spend 12 hours of class time on Part I. This part of the book gives all the basics for the different analyses that are carried out later. Therefore, we see it as mandatory for a teaching course. The chapters of Parts II and III can be studied more or less independently depending on the focus that lecturers want to set during their course. We suggest you spend four hours on each chapter of Parts II and III if they are made to be part of a course.

We thank all our colleagues who worked with us on bio-inspired computation during recent years. In particular, we like to mention the research groups at the University of Adelaide, Technical University of Berlin, University of Birmingham, Massachusetts Institute of Technology, Technical University of Denmark, Technical University of Dortmund, Max Planck Institute for Informatics, and Swiss Federal Institute of Technology Zurich.

Saarbrücken, Kongens Lyngby   *Frank Neumann*
September 2010   *Carsten Witt*

Contents

Part I

Basics

1

Introduction

Algorithms play an important role in computer science and are essential for several important applications. The term "algorithm" refers to a procedure to solve a given problem. Such a problem may have different features and structures, and in the case where the problem is well understood, specific algorithms may be designed that achieve good solutions for the problem at hand. The design and the analysis of such problem-specific algorithms has been widely studied for a wide range of problems (Cormen, Leiserson, Rivest, and Stein, 2001). The goal in this field of research is to obtain algorithms that are provably optimal with respect to the runtime and/or approximation ability for the studied problem. Studying a specific problem allows us to obtain knowledge about the problem at hand, which can be used for the development and the analysis of problem-specific algorithms. When looking at the results obtained in this field, the reader may observe the following. Often problem-specific algorithms are very complicated as they try to incorporate as much problem knowledge as possible so that good guarantees about the runtime and/or approximation quality can be proven. On the other hand, there are also many simple randomized algorithms available for which good performance guarantees can be given (Motwani and Raghavan, 1995). The proof that such simple algorithms work well is usually more complicated as knowledge about the problem is only implicitly present in the algorithm and is worked out in the analysis.

In many situations, it is not possible to develop problem-specific algorithms that have good performance guarantee. This is the case if a newly given problem is complex and/or has not been studied extensively before. In fact, most of the algorithms used in practice such as so-called bio-inspired computing and mixed integer programming do not come along with rigorous proofs that give bounds on the runtime and/or approximation quality. Instead they provide high performance results in experimental studies and it is often hard to understand why they perform well in a particular setting. Another important advantage is that these algorithms often can be applied without much

F. Neumann, C. Witt, *Bioinspired Computation*
in Combinatorial Optimization, Natural Computing Series,
DOI 10.1007/978-3-642-16544-3_1, © Springer-Verlag Berlin Heidelberg 2010

knowledge about the problem at hand, which makes them highly suitable for various applications.

We discuss such algorithms that are highly successful in practice. As their application often does not require specific knowledge about the problem at hand, we refer to them as general-purpose algorithms. The design of such an algorithm should only involve the following three steps, which makes it very easy to apply such methods to a newly given problem.

1. Choose a representation of possible solutions.
2. Determine a function to evaluate the quality of a solution.
3. Define operators that produce from a current set of solutions a new set of solutions.

Well-known simple approaches fitting this setting are local search (LS) (Aarts and Lenstra, 2003; Hoos and Stützle, 2004), and simulated annealing (SA) (van Laarhoven and Aarts, 1997). On the other hand, general-purpose algorithms have been designed that are inspired by processes observed in nature. Such algorithms belonging to the field of bio-inspired computation usually involve more complicated operators than the two approaches mentioned before. The field of bio-inspired computation covers many algorithmic approaches inspired by processes observed in nature. It includes well-known approaches such as evolutionary algorithms (EAs) (Eiben and Smith, 2007), ant colony optimization (ACO) (Dorigo and Stützle, 2004), and particle swarm optimization (PSO) (Kennedy and Eberhart, 1995).

Throughout this book, we will concentrate on evolutionary algorithms and ant colony optimization. However, we are optimistic that the insights and methods presented in this book may also be useful for obtaining similar results for other bio-inspired computation methods. Evolutionary algorithms are perhaps the most popular kind of algorithms belonging to the field of bio-inspired computation. They were introduced in the 1960s and have been applied to complex engineering problems as well as to problems from combinatorial optimization. In the case of a complex engineering problem, the structure of the problem is often not known. Then the quality of a certain parameter setting can often only be evaluated by experiments or simulations. Such problems are considered in the field of black-box optimization, where the value of a parameter setting can only be given after having executed some experiments or simulations. EAs have shown to be very successful on many problems from black-box optimization. In the case of combinatorial optimization, often much more is known about the structure of a given problem, and the function to be optimized can be given and analyzed. Nevertheless, it is often difficult to obtain good solutions for such problems, especially if the problem is new and there are not enough resources (such as time, knowledge, money) to design specific algorithms for the given problem.

Ant colony optimization is a more recent but also very well established branch in the field of bio-inspired computation, with the very first publications dating back to the early 1990s (Dorigo, Maniezzo, and Colorni, 1991).

This approach is inspired by the way ant colonies find shortest paths in unknown environments using pheromone trails as a means of communication. ACO algorithms construct solutions to problems by letting so-called artificial ants perform guided random walks on construction graphs. Promising paths through the graphs are rewarded by means of so-called pheromone updates which increase the probability of rediscovering and, possibly, further improving the solution. The underlying concept of a construction graph makes the approach very attractive for the solution of combinatorial optimization problems, in particular graph problems; however, in principle, any black-box problem in the above sense can be treated by ACO.

Due to the biological models that have been in mind when designing these algorithms, they are not designed for analysis in a classical sense. A major point in the design and analysis of algorithms is to prove bounds on the runtime that such algorithms have in order to obtain optimal or nearly optimal solutions. Often the design process for a certain problem is influenced by the goal of proving that the algorithm achieves good solutions quickly. In the case of bio-inspired computation methods, the goal was rather the development of algorithms that behave well for a wide range of problems by imitating optimization processes observed in nature. Such algorithms have then been examined experimentally to show their efficiency. Due to this background, bio-inspired computation methods are from a natural point of view not easy to analyze. However, there has been a lot of progress in understanding such methods rigorously in recent years. The goal of this book is to give an overview of the different results that have been achieved by studying the computational complexity of bio-inspired computation. We will, in particular, emphasize problems from combinatorial optimization as these algorithms have been used for many applications in this area.

Until the 1990s, theoretical work in the area of evolutionary algorithms was concentrated on showing that an algorithm converges to an optimal solution after a finite number of steps. In contrast, it has been considered what happens in one iteration of the algorithms. Although these are interesting investigations, the two aspects do not allow us to give upper or lower bounds on the runtime of an evolutionary algorithm for a considered problem.

As bio-inspired computation methods make use of a lot of random decisions, it seems appropriate to treat them as randomized algorithms in a classical sense. Therefore, we also refer to them as *stochastic search algorithms* to point out that we regard bio-inspired computation methods as algorithms that involve random decisions. Taking this point of view, it seems natural to analyze the runtime of stochastic search algorithms in a classical way, e.g., by bounding the expected runtime to achieve good solutions for a certain problem. Runtime analyses of bio-inspired computation methods have to consider problems where the function to be optimized can at least be covered analytically. As explained before, this is often not possible for complex engineering problems. Therefore, we consider combinatorial optimization problems, which seem to be natural, but non-trivial examples where bio-inspired computation methods have been applied. The rigorous analysis of such algorithms with respect to their runtime behavior is a relatively new research area. Most of the results in recent years have been obtained for evolutionary algorithms. Later on, newer variants such as ant colony optimization, particle swarm optimization, and artificial immune systems have been taken into account. We will mainly focus on evolutionary algorithms throughout this book but will discuss general methods that are also applicable for analyzing other bio-inspired computation methods.

The first theoretical result on the runtime of an evolutionary algorithm was given by Mühlenbein (1992). He presented an upper bound on the expected runtime of the simplest evolutionary algorithm, called (1+1) EA. The function considered is the simplest non-trivial pseudo-boolean function called OneMax, which counts the number of ones in a given bitstring. Since the mid-1990s, more rigorous results on the runtime of the (1+1) EA for different kinds of pseudo-boolean functions have been obtained. The first step was a much simpler proof for Mühlenbein's result and a generalization of the given bound to linear pseudo-boolean functions done by Droste, Jansen, and Wegener (2002). Considering different pseudo-boolean functions, the main aim was to show the behavior of EAs in different situations. Together with these results, many techniques have been developed that are very useful to analyze more complicated EAs as well as the behavior of bio-inspired computation methods on more natural problems. Recently, some of these techniques were transferred and further developed in order to prove the first rigorous results on the runtime of ACO (Gutjahr, 2007; Neumann and Witt, 2009) and PSO (Sudholt and Witt, 2010).

In this book, we present the major results that have been obtained regarding the computational complexity of bio-inspired computation methods for combinatorial optimization problems. We study some of the most prominent combinatorial optimization problems such as minimum spanning trees, Eulerian cycles, shortest paths, maximum matchings, scheduling and covering problems. We cannot hope that general-purpose algorithms beat the best-known problem-specific algorithms for such problems. Our goal is to understand which structures and problems can be provably solved efficiently by bio-inspired computation methods. Looking at the results presented in this book, we can observe that bio-inspired computation methods are efficient problem solvers for most of the mentioned problems although they only use a small amount of problem knowledge.

The book is divided into three major parts. The first one sets up and reviews concepts of stochastic search algorithms and tools for their analysis. Such analyses will be provided for two significant formulations of the investigated combinatorial problems, namely single-objective and multi-objective problems. The second part of the book concentrates on problems that are naturally formulated as single-objective ones. The third part elaborates on a classical multi-objective problem as well as on multi-objective reformulations of originally single-objective problems. This kind of reformulation may introduce helpful information into the stochastic search algorithms and provably speed them up compared to the plain single-objective problem formulation. Finally, the book contains an appendix with frequently used mathematical tools.

2

Combinatorial Optimization and Computational Complexity

Combinatorial optimization problems arise in several applications. Examples are the task of finding the shortest path from Paris to Rome in the road network of Europe or scheduling exams for given courses at a university. In this chapter, we give a basic introduction to the field of combinatorial optimization. Later on, we discuss how to measure the computational complexity of algorithms applied to these problems and point out some general limitations for solving difficult problems.

2.1 Combinatorial Optimization

Optimization problems can be divided naturally into two categories. The first category consists of problems with continuous variables. Such problems are well known from school courses on mathematics. A simple example consists of finding the minimum of the function $f : \mathbb{R} \to \mathbb{R}$ with $f(x) = x^2$. It is obvious that $x_0 = 0$ is the unique solution for this problem. More complicated problems are often tackled by computing the derivatives, using Newton methods or linear programming techniques. As this book deals with combinatorial optimization problems, we will not go into detail the different methods to tackle continuous optimization problems, and refer the interested reader to Nocedal and Wright (2000).

In the case of discrete variables we are dealing with discrete optimization. When speaking of combinatorial optimization problems, most people have "natural" discrete optimization problems in mind, such as computing shortest paths or scheduling different jobs on a set of available machines. In a combinatorial optimization problem, one aims at either minimizing or maximizing a given objective function under a given set of constraints.

A problem consists of a general question that has to be answered and is given by a set of input parameters. An instance of a problem is given by the problem together with a specified parameter setting. Formally, a combinatorial optimization problem can be defined as a triple (S, f, Ω), where S is a given

F. Neumann, C. Witt, *Bioinspired Computation*
in Combinatorial Optimization, Natural Computing Series,
DOI 10.1007/978-3-642-16544-3_2, © Springer-Verlag Berlin Heidelberg 2010

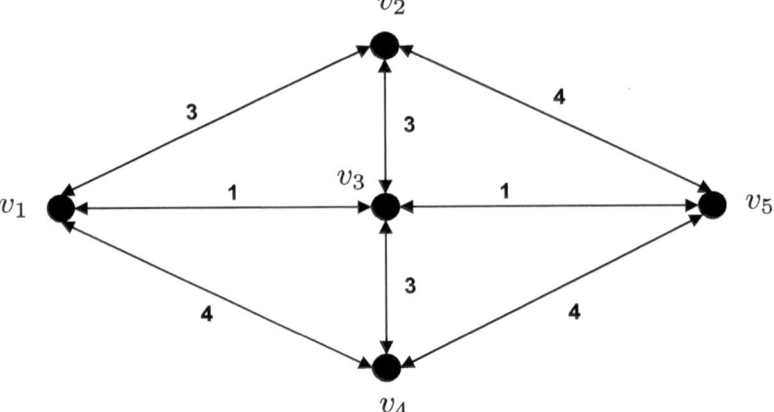

Fig. 2.1. Example graph G

search space, f is the objective function, which should be either maximized or minimized, and Ω is the set of constraints that have to be fulfilled to obtain feasible solutions. The goal is to find a globally optimal solution, which is in the case of a maximization problem a solution s^* with the highest objective value that fulfills all constraints. Similarly, in the case of minimization problems, one tries to achieve a smallest objective value under the condition that all constraints are fulfilled.

Throughout this book, we consider many combinatorial optimization problems on graphs. A directed graph G is a pair $G = (V, E)$, where V is a finite set and E is a binary relation on V. The elements of V are called vertices. E is called the edge set of G and its elements are called edges. For an illustration see Figure 2.1.

We use the notation $e = (u, v)$ for an edge in a directed graph. Note that self-loops that are edges of the kind (u, u) are possible. In an undirected graph $G = (V, E)$, no self-loops are possible. The edge set E consists of unordered pairs of vertices in this case, and an edge is a set $\{u, v\}$ consisting of two distinct vertices $u, v \in V$. Note that one can think of an undirected edge $\{u, v\}$ as two directed edges (u, v) and (v, u). If (u, v) is an edge in a directed graph $G = (V, E)$ we say that v is adjacent to vertex u. This leads to the representation of graphs by adjacency matrices, which will be discussed later in greater detail. A path of length k from a vertex v_0 to a vertex v_k in a graph $G = (V, E)$ is a sequence $v_0, v_1, \ldots, v_k$ of vertices such that $(v_{i-1}, v_i) \in E$, $1 \leq i \leq n$, holds. Note that a path implies a sequence of directed edges. Therefore, it is sometimes useful to denote a path $(v_0, v_1, \ldots, v_k)$ by its sequence of directed edges $(v_0, v_1), (v_1, v_2), \ldots, (v_{k-1}, v_k)$.

The graph G in Figure 2.1 consists of the vertex set

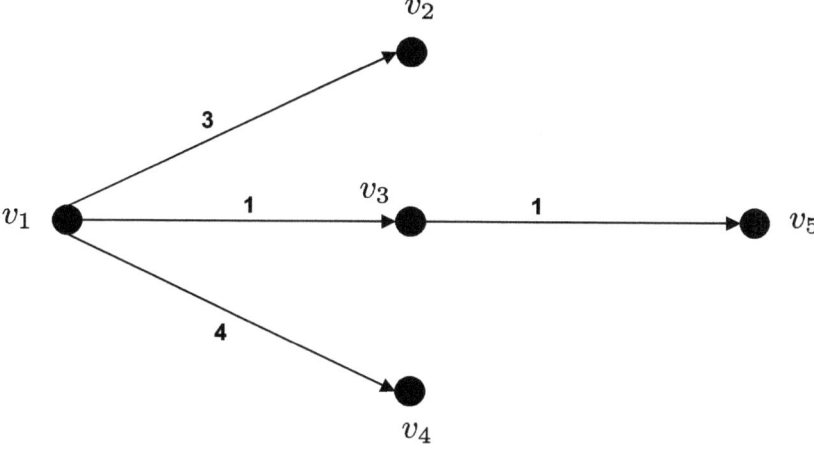

Fig. 2.2. Single source shortest path tree for G and $s = v_1$

$$V = \{v_1, v_2, v_2, v_3, v_4, v_5\}$$

and the edge set

$$E = \{e_1, e_2, e_3, e_4, e_5, e_6, e_6, e_7, e_8\}$$

where $e_1 = \{v_1, v_2\}, e_2 = \{v_1, v_3\}, e_3 = \{v_1, v_4\}, e_4 = \{v_2, v_3\}, e_5 = \{v_2, v_5\}$, $e_6 = \{v_3, v_4\}, e_7 = \{v_3, v_5\}$, and $e_8 = \{v_4, v_5\}$. In addition, there is a weight function $w\colon E \to \mathbb{N}$ assigning weights to the edges, i.e., $w(e_1) = w(e_4) = w(e_6) = 3$, $w(e_2) = w(e_7) = 1$, and $w(e_3) = w(e_5) = w(e_8) = 4$. Clearly, (v_1, v_2, v_3, v_5) is a path in G whereas (v_1, v_5, v_2) is not as there is no edge from v_1 to v_5.

There are many well-known combinatorial optimization problems on weighted graphs. We want to introduce two basic problems in the following. In the case of the single source shortest path problem, an undirected connected graph $G = (V, E)$ with positive weights on the edges is given. The goal is to compute from a designated vertex $s \in V$ the shortest paths to all other vertices of $V \setminus \{s\}$. The solution of this problem can be given by a tree rooted at s which contains the shortest paths. Considering the graph G of Figure 2.1 and $s = v_1$, a shortest path tree is shown in Figure 2.2. Another well-known combinatorial optimization problem on undirected connected graphs with positive weights is the minimum spanning tree problem. Here, one searches for a connected subgraph of the given graph G that has minimal cost. As the edge weights are positive, such a graph does not contain cycles, i.e., it is a tree. Considering again the graph G of Figure 2.1, a minimum spanning tree of G is given in Figure 2.3.

Other important problems on graphs are covering problems. In the case of the so-called vertex cover problem for a given undirected graph $G = (V, E)$,

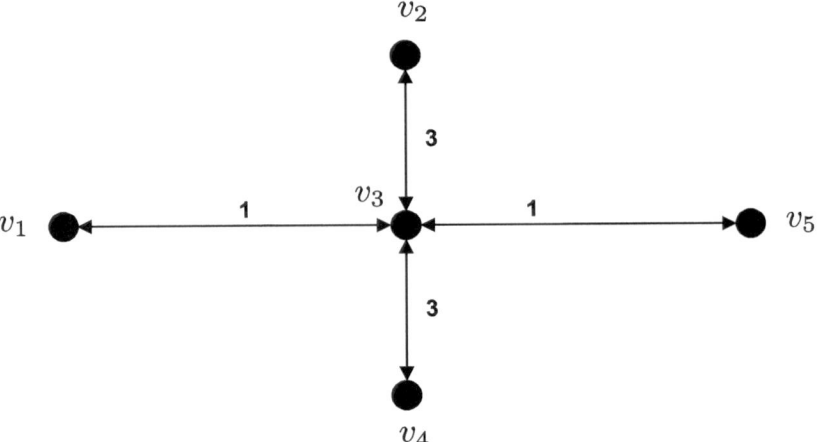

Fig. 2.3. Minimum spanning tree of G

one searches for a minimal subset of vertices $V' \subseteq V$ such that each edge $e \in E$ contains at least one vertex of V, i.e., $\forall e \in E: e \cap V' \neq \emptyset$ holds.

Another class of combinatorial optimization problems that has been widely examined in the literature is scheduling problems. Here, n jobs are given that have to be processed on $m \geq 1$ machines. Associated with each job j, $1 \leq j \leq n$, is usually a processing time p_j. The processing time need not be the same for each machine. There are variants of scheduling problems where the processing time may depend on the machine by which it is processed. Often, also a specific due date for each job is given. Consider the following simple scheduling problem on two machines. Given are n jobs and for each job j a processing time p_j which holds independently of the chosen machine. The goal is to find an assignment of the jobs to the two machines such that the overall completion time is minimized. Let $x \in \{0,1\}^n$ be a decision vector. Job j is on machine 1 iff $x_j = 0$ holds and on machine 2 iff $x_j = 1$ holds. The goal is to minimize

$$\max \left\{ \sum_{i=1}^{n} p_j x_j, \sum_{i=1}^{n} p_j (1 - x_j) \right\}.$$

2.2 Computational Complexity

In contrast to the description of a problem, which is usually short, the search space is most of the time exponential in the problem dimension. In addition, for a lot of combinatorial optimization problems, one cannot hope to come up with an algorithm that produces for all problem instances an optimal solution within a time bound that is polynomial in the problem dimension. The performance measure most widely used to analyze algorithms is the time an algorithm takes to present its final answer. Time is expressed in terms of number of elementary operations such as comparisons or branching instructions (Papadimitriou and Steiglitz, 1998). The time an algorithm needs to give the final answer is analyzed with respect to the input size. The input of a combinatorial optimization problem is often a graph or a set of integers. This input has to be represented as a sequence of symbols of a finite alphabet. The size of the input is the length of this sequence, that is, the number of symbols in it.

In this book, we are dealing with combinatorial optimization problems. Often we are considering a graph $G = (V, E)$ with n vertices and m edges and are searching for a subgraph $G' = (V', E')$ of the given one that fulfills given properties.

One approach to represent a graph is to do it by an adjacency matrix $A_G = [a_{ij}]$, where $a_{ij} = 1$ if $(v_i, v_j) \in E$ and $a_{ij} = 0$ otherwise. This matrix has n^2 entries, i.e., the number of entries is quadratic with respect to the number of vertices. An entry $a_{ij} = 1$ means that there is an edge from v_i to v_j and $a_{ij} = 0$ holds if this is not the case. Note that the adjacency matrix of a given undirected graph is symmetric. An undirected graph may have up to $\binom{n}{2} = \Theta(n^2)$ edges. However, if we are considering so-called sparse graphs, the number of edges is far less than $\binom{n}{2}$.

In the case of sparse graphs, it is better to represent a given graph by so-called adjacency lists. Here, for each vertex $v \in V$ we record a set $A(v) \subseteq V$ of vertices that are adjacent to it. The size of the representation is given by the sum of the length of lists. As each edge contributes 2 to this total length, we have to write down $2m$ elements. Another factor which effects the total length of the representation is how to encode the vertices. Our alphabet has finite size. Assume the alphabet is the set $\{0, 1\}$. Therefore we need $\Theta(\log n)$ bits to encode one single vertex. This implies that we need $\Theta(m \log n)$ bits (or symbols) to represent the graph G. In practice we say that a graph G can be encoded in $\Theta(m)$ space, which seems to be a contradiction to the previous explanation. The reason is that computers treat all integers in their range the same. Here the same space is needed to store small integers such as 5 or large integers such as 3^{12}. We assume that graphs are considered where the number of vertices is within the integer range of the computer. This range is in most cases 0 to 2^{31}, which means that integers are represented by 32 bits. Therefore $\Theta(m)$ is a reasonable approximation of the size of a graph and analyzing graph algorithms with respect to m is accepted in practice. In most cases both parameters n and m are taken into account when analyzing the complexity of a graph algorithm.

Considering graph algorithms where we can bound the runtime by a polynomial in n and m, we obviously get a polynomial-time algorithm. We have to be careful when the input includes numbers. Let $N(I)$ be the largest integer that appears in the input. An algorithm A is called pseudo-polynomial if it is polynomial in the input size $|I|$ and $N(I)$. Note that $N(I)$ can be encoded by $\Theta(\log(N(I)))$ bits. Therefore a function that is polynomial in $|I|$ and $N(I)$ is not necessarily polynomial in the input size. Often the input consists of small integers. In the case where $N(I)$ is bounded by a polynomial in $|I|$, A is a polynomial-time algorithm.

An important issue that comes up when considering combinatorial optimization problems is the classification of difficult problems (Papadimitriou and Steiglitz, 1998). To distinguish between easy and difficult problems, one considers the class of problems that are solvable by a deterministic Turing machine in polynomial time and problems that are solvable by a nondeterministic Turing machine in polynomial time. We do not want to formalize the characterization of the classes P and NP via Turing machines and prefer to outline the characteristics and notions connected with these classes at a more intuitive level. This leads to a straightforward definition to characterize problems that belong to P.

Definition 2.1. *A problem is in P iff it can be solved by an algorithm in polynomial time.*

Problems in P can therefore be solved in polynomial time by using an appropriate algorithm. Examples of problems belonging to this class are the single source shortest path problem and the minimum spanning tree problem introduced in Section 2.1.

A class that is intuitively associated with hard problems is called NP. Typically, NP is restricted to so-called decision problems, i.e., problems whose output is either YES or NO. This restriction has a technical background and captures the essentials of the problems without simplifying them too much.

Definition 2.2. *A decision problem is in NP iff any given solution of the problem can be verified in polynomial time.*

For problems in NP, it is therefore not necessary that a solution be computable in polynomial time. It is only necessary that we can verify the solution of the problem in polynomial time. Therefore $P \subseteq NP$ holds (slightly abusing notation by restricting P to decision problems), and it is widely assumed that $P \neq NP$.

Consider the following decision variant of the vertex cover problem. The question is whether a given graph $G = (V, E)$ contains a vertex cover of at most k vertices. Given a solution x we can easily check whether each edge is covered by x. This can be done in linear time by examining each edge at most once. Additionally, we can count the number of vertices chosen by x in linear time and therefore verify whether x is a vertex cover with at most k vertices in polynomial time.

Many optimization and decision problems, including the vertex cover problem, are at least as difficult as any problem in NP. Such problems are called

NP-hard. Showing that a problem is *NP*-hard is usually done by giving a polynomial-time reduction from an *NP*-hard problem to the considered problem. This reduction involves a transformation of the known *NP*-hard problem to the considered one, which has to be done in polynomial time. Such a reduction links the considered problem to the known *NP*-hard problem in such a way that iff the considered problem can be solved in polynomial time also the *NP*-hard problem to which it has been reduced can. We do not want to go into the details and refer the reader to a book on complexity theory (Wegener, 2005a) for further reading.

Definition 2.3. *A problem is called NP-hard iff it is at least as difficult as any problem in NP, i.e., each problem in NP can be reduced to it.*

As we are considering optimization problems in this book, we want to point out that many optimization problems are *NP*-hard but not in *NP*. We consider the vertex cover problem again, but at this time its optimization variant where the task is to compute a vertex cover of minimal size. Clearly, this optimization variant is at least as difficult as the problem of deciding whether a given graph contains a vertex cover of at most k vertices. However, since the output of the optimization problem is a number, it is not a decision problem and, therefore, not in *NP*.

In summary, many optimization problems are at least as difficult as any problem in *NP*, i.e., *NP*-hard but not in *NP*. Problems that are *NP*-hard and also in *NP* are called *NP*-complete. This holds for many decision variants of *NP*-hard optimization problems.

Definition 2.4. *A problem is NP-complete iff it is NP-hard and in NP.*

The classical approach to deal with *NP*-hard problems is to search for good approximation algorithms (Hochbaum, 1997; Vazirani, 2001). These are algorithms that run in polynomial time but guarantee that the produced solution is within a given ratio of an optimal one. Such approximation algorithms can be totally different for different optimization problems. In the case of the *NP*-hard bin packing problem, even simple greedy heuristics work very well whereas in the case of more complicated scheduling problems often methods based on linear programming are used.

Another approach to solve *NP*-hard problems is to use sophisticated exact methods that have in the worst case an exponential runtime. The hope is that such algorithms produce good results for interesting problem instances in a small amount of time. A class of algorithms that tries to come up with exact solutions is branch and bound. Here the search space is shrunk during the optimization process by computing lower bounds on the value of an optimal solution in the case where we are considering maximization problems. The hope is to come up in a short period of time with a solution that matches such a lower bound. In this case an optimal solution has been obtained.

Related to this is the research on *parametrized complexity* (Downey and Fellows, 1999). Here, parametrized versions of given optimization problems are studied. These are usually decision problems in the classical sense. Consider for example the decision variant of the vertex cover problem where we ask whether a given graph has a vertex cover of at most k vertices. This question can be answered in time $O(1.2738^k + kn)$ (Chen, Kanj, and Xia, 2006), i.e., in polynomial time for any fixed k, and a corresponding solution with k vertices can be computed within that time bound if it exists. Obviously, this approach can be turned into an optimization algorithm that is efficient iff the value of an optimal solution is small.

A crucial consideration in combinatorial optimization problems and stochastic search algorithms that search more or less locally is the neighborhood of the current search point. Let $s \in S$ be a search point in a given search space. The neighborhood is defined by a mapping $N \colon S \to 2^S$. In the case we are considering combinatorial optimization problems from the search space $\{0, 1\}^n$, the neighborhood can be naturally defined by all solutions having at most Hamming distance k from the current solution s. The parameter k determines the size of the neighborhood from which the next solution is sampled. Choosing a small value k, e.g. $k = 1$, such a heuristic may get stuck in local optima. If the value of k is large (in the extreme case $k = n$) and all search points of the neighborhood are chosen with the same probability, the next solution will be somehow independent of s. This leads to stochastic search algorithms that behave almost as if they were choosing in each step a search point uniformly at random from $\{0, 1\}^n$. In this case the stochastic search algorithm does not take the previously sampled function values into account and the search cannot be directed into "good" regions of the considered search space.

2.3 Approximation Versus Exact Optimization

As already mentioned, NP-hard problems probably do not allow exact solutions in polynomial time, so good approximations of optimal solutions are desired. A formal definition of the quality of approximations is based on a fixed approximation algorithm and the worst case from the set of instances for the combinatorial optimization problem.

Definition 2.5. *Given an algorithm A for the solution of a combinatorial optimization problem (S, f, Ω), let $s_A \in S$ denote a solution produced by A and $f_A := f(s_A)$ its f-value. Given f_{opt}, the f-value of an optimal solution, the approximation ratio of s_A is defined by f_A/f_{opt} for minimization problems and by f_{opt}/f_A for maximization problems.*

We say that an algorithm maintains a certain approximation ratio if it produces solutions of this approximation ratio on all instances of the underlying problem. In particular, we are interested in algorithms achieving a certain approximation ratio within polynomial time.

Definition 2.6. *A polynomial-time approximation algorithm with ratio r to a combinatorial optimization problem is an algorithm that computes solutions of approximation ratio r in polynomial time with respect to the input size.*

In the previous definition, r might depend on the problem size, which is for example the case if the possible approximation ratios become worse for growing inputs. The special case of a *constant* approximation ratio is given if r can be bounded independently of the problem size. Often, constant approximation ratios are obtainable even for *NP*-hard problems. An even stronger property is demanded by specifying the constant approximation ratio as a parameter of the approximation algorithm.

Definition 2.7. *A polynomial-time approximation scheme (PTAS) to a combinatorial optimization problem is an algorithm with parameter ϵ that computes solutions of approximation ratio $1 + \epsilon$ in polynomial time with respect to the input size. If the time is also polynomial with respect to $1/\epsilon$, the algorithm is called fully polynomial-time approximation scheme (FPTAS).*

Definitions 2.6 and 2.7 require polynomial time with probability 1 and are more suitable for deterministic than for randomized algorithms. A natural relaxation of the definitions is to allow expected polynomial time, resulting in expected-polynomial-time approximation algorithms and schemes. However, it is more convenient to prescribe polynomial time with a certain success probability. This results in the following definition (Motwani and Raghavan, 1995).

Definition 2.8. *A polynomial-time randomized approximation scheme (PRAS) to a combinatorial optimization problem is an algorithm with parameter ϵ that with probability at least 3/4 computes solutions of approximation ratio $1 + \epsilon$ in polynomial time with respect to the input size.*

The somewhat mysterious bound 3/4 on the success probability goes back to applications of PRASs to a generalization of optimization problems, the so-called number problems. However, the exact value is not too significant. Any constant success probability can be boosted to at least 3/4 by running the approximation algorithm a constant number of times and taking the best solution out of the runs. In the domain of EAs, this is usually referred to as *multistart* schemes.

We will get to know characterizations of EAs as approximation algorithms in Chapter 12 and characterizations as PRASs in Chapters 6 and 7.

2.4 Multi-objective Optimization

Many problems in computer science ask for solutions with certain attributes or properties that can be expressed as functions mapping possible solutions to scalar numeric values. The usual optimization approach is to take these attributes as constraints to determine the feasibility of a solution, while one of them is chosen as an objective function to determine the preference order of the feasible solutions. In the minimum spanning tree problem, as a simple example, constraints are imposed on the number of connected components (one) and the number of cycles (zero) of the chosen subgraph, while the total weight of its edges is the objective to be minimized.

A more general approach is multi-objective optimization (Ehrgott, 2005), where several attributes are employed as objective functions and used to define a partial preference order of the solutions, with respect to which the set of minimal (maximal) elements is sought. Most of the best known single-objective polynomial solvable problems like shortest path or minimum spanning tree become NP-hard when at least two weight functions have to be optimized at the same time. In this sense, multi-objective optimization is considered as more (at least as) difficult than (as) single-objective optimization.

In the case of multi-objective optimization, the objective function $f = (f_1, \ldots, f_k)$ is vector-valued, i.e., $f \colon S \to \mathbb{R}^k$. Since there is no canonical complete order on $\mathbb{R}^k$, one compares the quality of search points with respect to the canonical partial order on $\mathbb{R}^k$, namely $f(s) \leq f(s')$ iff $f_i(s) \leq f_i(s')$ for all $i \in \{1, \ldots, k\}$. A Pareto optimal search point s is a search point such that (in the case of minimization problems) $f(s)$ is minimal with respect to this partial order and all $f(s'), s' \in S$. Again, there can be many Pareto optimal search points, but they do not necessarily have the same objective vector. The Pareto front, denoted by F, consists of all objective vectors $y = (y_1, \ldots, y_k)$ such that there exists a search point s where $f(s) = y$ and $f(s') \leq f(s)$ implies $f(s') = f(s)$. The Pareto set consists of all solutions whose objective vector belongs to the Pareto front. The problem is to compute the Pareto front and for each element y of the Pareto front one search point s such that $f(s) = y$. We sometimes say that a search point s belongs to the Pareto front, which means that its objective vector belongs to the Pareto front.

As in the case of optimization problems, one may be satisfied with approximate solutions. This can be formalized as follows. For each element y of the Pareto front, we have to compute a solution s such that $f(s)$ is close enough to y. Close enough is measured by an appropriate metric and an approximation parameter. In the single-objective case, one switches to the approximation variant if exact optimization is too difficult. The same reason may hold in the multi-objective case. There may be another reason. The size of the Pareto front may be too large for exact optimization.

The Pareto front F may contain exponentially many objective vectors. Papadimitriou and Yannakakis (2000) have examined how to approximate the Pareto front for different multi-objective combinatorial optimization problems. W. l. o. g., they have considered the task of maximizing all objective functions. Given an instance I and a parameter $\epsilon > 0$ they have examined how to obtain an ϵ-approximate Pareto set. This is a set of solutions X with the property that there is no solution s' such that for all $s \in X$ $f_i(s') \geq (1 + \epsilon) \cdot f_i(s)$

holds for at least one i. Papadimitriou and Yannakakis (2000) showed that there exists an algorithm which constructs such a set X, which is polynomially bounded in $|I|$ and $1/\epsilon$ if and only if the corresponding gap problem problem can be solved. Given an instance I of the considered problem and a vector $(b_1, \ldots, b_k)$, the gap problem consists of either presenting a solution s with $f_i(s) \geq b_i$, $1 \leq i \leq k$, or answering that there is no solution s' with $f_i(s') \geq (1+\epsilon) \cdot b_i$, $1 \leq i \leq k$. In the case of some multi-objective optimization problems (e.g., the multi-objective variants of the minimum spanning tree problem and the shortest path problem), such a set can also be computed within a time bound that is polynomial in $|I|$ and $1/\epsilon$. Algorithms with such properties constitute an FPTAS (Definition 2.7), which is the best we can hope for when dealing with NP-hard problems.

Stochastic Search Algorithms

We want to analyze bio-inspired computation methods in a rigorous way with respect to their runtime behavior. As these algorithms make use of many random decisions, we treat them as randomized algorithms to study their behavior in a rigorous manner. The term *stochastic search algorithms* stresses this point of view and will be used in the following to point out that bio-inspired computation methods can be treated as algorithms which are based on random decisions. Mainly we will consider stochastic search algorithms belonging to the field of evolutionary computation throughout this book. These algorithms are inspired by the evolution process in nature and follow Darwin's principle of the survival of the fittest. We take a closer look at the different approaches developed in this field in Section 3.1. Another kind of bio-inspired stochastic search algorithm is ant colony optimization, which will be introduced in Section 3.2. Here, solutions for a given problem are constructed by walks of ants on a so-called construction graph. To give a more complete picture, we describe other popular variants in Section 3.3.

A stochastic search algorithm is a problem-independent algorithm to solve problems from a considered search space although it might have modules that are adjusted to the considered problem or are combined with problem-dependent algorithms. The independence from the considered problem distinguishes stochastic search algorithms from problem-dependent algorithms developed and analyzed in the classical algorithm community. In contrast to the classical approach to algorithms, where one designs an algorithm with the task to prove bounds on the runtime and/or approximation quality in mind, stochastic search algorithms are general-purpose algorithms. Assuming that one considers different problems from the same search space, e.g., $\{0,1\}^n$, a stochastic search algorithm is usually applicable to each of these problems. Their easy adaptation to different problems usually has to be paid for by the disadvantage that the algorithm is often not rigorously analyzed with respect to its runtime and/or approximation quality.

In the general approach, the only problem-dependent component of the algorithm is the fitness function that guides the search. This function is the

F. Neumann, C. Witt, *Bioinspired Computation*
in Combinatorial Optimization, Natural Computing Series,
DOI 10.1007/978-3-642-16544-3_3, © Springer-Verlag Berlin Heidelberg 2010

only part of such an algorithm that has to be adjusted to the considered problem. Therefore, we get algorithms that can be implemented very easily and adjusted quickly to similar problems. As already mentioned, stochastic search algorithms are not designed with the focus of proving special properties on the runtime or approximation quality. This makes a rigorous analysis of such algorithms more difficult than the analysis of algorithms that have been designed in a special way to prove properties such as the runtime or approximation quality of the algorithm.

We start with a general description of stochastic search algorithms, which covers all important approaches such as evolutionary algorithms, ant colony optimization, randomized local search, the Metropolis algorithm, and simulated annealing.

Given a search space S, the aim is to optimize a considered function $f \colon S \to R$, where R is the set of all possible function values. A stochastic search algorithm working in a given search space S under the consideration of a function f chooses the first search point s_1 with respect to a probability distribution on S that may be determined by a heuristic. After that the function value $f(s_1)$ is computed. The search point s_t is chosen according to a probability distribution that can depend on the previous sampled search points $s_1, \ldots, s_{t-1}$ and their function values. The process is iterated until a stopping criterion has been fulfilled.

The No Free Lunch Theorem by Wolpert and Macready (1997) shows the basic limitations of stochastic search algorithms when considering the optimization of all possible functions. It is assumed that each search point of the considered search space is not evaluated more than once. This is a realistic restriction as function values for evaluated search points can be stored such that another evaluation is not necessary. Wolpert and Macready (1997) have given the following result.

Theorem 3.1 (No Free Lunch (NFL) Theorem). *Let S and R be two finite sets, $F = R^S$ be the set of all functions $f \colon S \to R$, and A and A' be two stochastic search algorithms that do not evaluate each search point more than once. Then the average number of fitness evaluations among all functions of F is the same for A and A'.*

This implies that no stochastic search algorithm behaves on the average better than blind random search, where in each step a solution is drawn uniformly at random from the so far unseen part of the search space. This should make clear that an analysis of these algorithms with respect to their runtime makes sense only for specific classes of functions or specific classes of problems.

3.1 Evolutionary Algorithms

Evolutionary algorithms (EAs) have become quite popular since the mid-1960s. Many different approaches have been proposed in the last 40 years. In this section, we give a brief overview of the main approaches proposed in the literature. For a more complete overview we refer the reader to a general book on evolutionary computation (Eiben and Smith, 2007).

Inspired by the evolution process in nature, evolutionary algorithms try to solve problems by evolving sets of search points such that satisfying results are obtained. A lot of the tasks that have been solved by EAs lie in the field of real-world applications. In real-world applications, the function to be optimized is often unknown and function values can only be obtained by experiments. Often these experiments have high costs or need a large amount of time. Therefore, the main aim is to minimize the number of function evaluations until a satisfying result has been obtained.

The main difference between evolutionary algorithms and local search procedures or simulated annealing is that evolutionary algorithms usually work at each time step with a set of solutions which is called the population of an EA. This population produces a set of solutions, called the offspring population, by some variation operators such as crossover or mutation. After that, a new population is created by selecting individuals from the parent and offspring population as a result of the fitness function f. We consider discrete search spaces throughout this book. In this case another important issue is that evolutionary algorithms often have a positive probability of sampling each search point of the given search space in the next step. In the case of local search and simulated annealing, this is usually not the case. There, the search points that can be constructed in the next step depend on the current solution and the neighborhood defined for the search process. Especially in the case of multi-objective optimization, EAs seem to be a good heuristic approach to obtain a good set of solutions. EAs have the advantage that their population may be evolved to obtain a good approximation of the Pareto front.

3.1.1 Representation

We want to take a look at the different modules of an EA. The first important issue is representation. Solutions can be represented in different ways. A good example is the different representations of spanning trees. For a given undirected connected graph with n vertices and m edges, the most natural representation seems to be a set of $n - 1$ edges such that the graph is connected. This is known as the representation of spanning trees by edge sets (Raidl and Julstrom, 2003). It is more general to represent them as bitstrings of length m, where each bit corresponds to an edge which is included in the solution if the bit is set to 1 and excluded otherwise. In this case, information to obtain a connected graph or a spanning tree has to be incorporated into the fitness function. We will use this representation later for the analysis of EAs on the minimum spanning tree problem. There, it turns out that guiding such an algorithm to compute connected graphs or spanning trees by the fitness function is a minor term in the overall complexity.

Spanning trees can also be represented by Prüfer numbers. A Prüfer number consists of $n-2$ node identifiers which determine a spanning tree. This number can be decoded by an algorithm into a corresponding spanning tree and a spanning tree can be encoded into a Prüfer number using a complementary algorithm. The disadvantage is that small changes in the Prüfer number can result in a totally different spanning tree. Therefore Prüfer numbers are a poor representation of spanning trees when using an EA (Gottlieb, Julstrom, Raidl, and Rothlauf, 2001). This is not the case when edge sets are considered. If the set of edges is changed by one edge, then the two spanning trees have $n-2$ edges in common. It should be clear that this point of locality is important for the success of an EA. If only small changes lead to a completely different solution with a fitness value that does not depend on the last sampled search point, the search cannot be directed into "good" regions of the search space.

3.1.2 Variation Operators

Variation operators are important for constructing new solutions. They have to be adjusted to the chosen representation. The most popular variation operators are mutation and crossover. In the case of mutation, one single individual is altered, in a crossover operation at least two individuals produce new solutions. Often, in a first step crossover is used to produce offspring and these offspring are additionally altered by a mutation operator. Throughout this book, we will analyze EAs that use only mutation to obtain new solutions.

Nevertheless, to give a more complete picture, we also present some popular crossover operators for the search space $\{0,1\}^n$ and the representation of permutations. Crossover operators produce new search points by combining search points of the current population. We first take a look at the case where solutions are represented as bitstrings of length n. The most important crossover operators for individuals that are bitstrings of length n are uniform and k-point crossover, where usually $k \in \{1,2\}$ is chosen. Consider two individuals $x = (x_1, \ldots, x_n)$ and $y = (y_1, \ldots, y_n)$ that should produce a new solution $z = (z_1, \ldots, z_n)$ by a crossover operator. In the case of uniform crossover $\mathrm{Prob}(z_i = x_i) = \mathrm{Prob}(z_i = y_i) = 1/2$ if $x_i \neq y_i$ holds. Otherwise $z_i = x_i = y_i$ holds for the created child z. In the case of k-point crossover, k positions in the two bitstrings are selected at random. Based on these positions the individuals are partitioned into different intervals, where the intervals are numbered based on their position in the bitstrings. The new individual z is formed by taking all entries of intervals with odd numbers from x and all entries of intervals with even numbers from y.

In the case of the representation of permutations, it is a little bit more difficult to obtain sensible crossover operators. We assume that we are working with permutations consisting of n elements. Most crossover operators are applied to two parents P_1 and P_2 and produce two offspring O_1 and O_2. To give an impression of how crossover operators for permutation problems are designed we consider the order crossover operator (OX-operator), which gets two parameters i and j, $1 \le i, j \le n$. W.l.o.g., we assume $i < j$. In a first step the elements of P_1 at positions $i + 1, \ldots, j - 1$ are copied into O_1 to the same positions. After that the remaining elements of P_2 are placed into O_1. This is done by examining P_2 from position j on in a circular way and placing the elements that up to now do not occur in O_1 at the next position, where the positions $j, \ldots, n, 1, \ldots, i$ are considered one after another. In the same way the offspring O_2 is constructed by starting copy the elements between the positions i and j of P_2 into O_2.

We describe important mutation operators for the search space of binary strings and permutations of elements in the following. In the case of bitstrings of length n each bit is often flipped with a certain probability p, where $p = o(1)$ usually holds. It is necessary to choose p not too large to prevent the algorithm from sampling the next solution nearly uniformly at random from a very large neighborhood of the parent solution. In a lot of algorithms $p = 1/n$ is used such that on average one bit is flipped. In the case of permutations with n elements, often jumps or exchange operations are used. Both operations get two parameters i and j, $1 \le i, j \le n$. Then a $jump(i, j)$ places the element at position i at position j and shifts the elements between i and j, including j, in the appropriate direction. If $i < j$ the elements are shifted to the left, and to the right if $i > j$. An $exchange(i, j)$ places the element at position i at position j and the element at position j at position i. W.l.o.g., assume that $i < j$ holds for $exchange(i, j)$. Then this operation can be simulated by executing sequentially the two jump operations $jump(i, j)$ and $jump(j-1, i)$. In contrast to this, $\lfloor k/2 \rfloor$ exchange operations are needed to simulate $jump(i, j)$ if $|i - j| = k$ holds. Therefore the jump operator seems to be the more flexible one. We will see later that this can make the difference between a polynomial and an exponential expected runtime.

3.1.3 Selection Methods

Selection methods are used to decide which individuals of the current population are used to produce offsprings. In addition, they are used to decide which individuals from the parent and offspring population constitute the population of the next generation. A widely used selection method is fitness-proportional selection. We assume that the function f should be maximized and that all function values are positive. If the population contains μ individuals $x^1, \ldots, x^\mu$, then x^i has probability $f(x^i)/(\sum_{i=1}^{\mu} f(x^i))$ of being chosen in each selection step. Note that this selection method allows us to choose individuals more than once for a certain purpose. Therefore the population of the next generation may include duplicates even if the parent and offspring population before have contained only individuals that were pairwise distinct from each other.

Another important method is tournament selection. Here, tournaments of size $q \in \{1, \ldots, \mu\}$ are chosen. In each tournament, q individuals compete against each other. The individuals that take part in a certain tournament are chosen uniformly at random from the population. In each tournament, the individual with the highest fitness value wins the competition and is chosen for reproduction for the next generation.

Two other important selection methods are $(\mu + \lambda)$- and (μ, λ)-selection. These two methods have their main application in evolution strategies. We will discuss the different approaches in evolutionary computation together with these two methods in the following.

3.1.4 Major Approaches

The class of evolutionary algorithms covers historically different approaches to solve problems inspired by the evolution process in nature. The approaches differ by the search spaces that are considered and the variation operators used to produce new search points.

Evolution Strategies

Evolution strategies (ESs) (Rechenberg, 1973; Schwefel, 1981) are used to solve continuous optimization problems. There, usually a real-valued search space is considered. Mutation is the variation operator that is mainly used in ES. The most important strategies are called (μ, λ)- and $(\mu + \lambda)$-ES and differ from each other by the chosen selection method. In the case of a (μ, λ)-ES, the parent population has size μ and λ children are produced in one generation. The next parent population is created by choosing μ individuals from the offspring population. Note that in this case $\lambda >> \mu$ should hold as the parent population is not involved in the selection process. In contrast to this, a $(\mu + \lambda)$ strategy considers both populations for the next parent population. After having created λ children, individuals from the parent and the offspring population are chosen according to their fitness values to build the parent population of the next generation.

Genetic Algorithms

Genetic algorithms (GAs) introduced by Holland (1975), work in discrete search spaces. Here, bitstrings of length n are used to represent possible solutions. The other main difference with evolution strategies is that crossover is seen as the variation operator that has the main effect of getting good solutions. Working with a population of size μ, in each iteration μ children are produced by using crossover. Mutation is seen as the minor variation operator. If it takes place, it is often applied to each child that has been produced by crossover. Then each bit is flipped with a certain probability p, where often $p = 1/n$ is chosen. The major selection method for GAs is fitness proportional selection. This method is used to select the individuals that are used to obtain new solutions as well as to select the individuals from the parents and children to form the population of the next generation. Another variant is to produce only a few children in each iteration. In the extreme case, one child is produced. This is known as the steady state GA. A lot of theoretical work for GAs has been concentrated on schemata. A schema fixes some positions in the bitstrings such that a search space of a smaller dimension is obtained. It is assumed that genetic algorithms combine schemata to obtain better ones. This implicitly assumes that the function which should be optimized is separable and comforms to the so-called building block hypothesis. This hypothesis says that functions are optimized by separating the variables and optimizing functions that depend on these partitions. It is assumed that such a partitioning is found by a GA and that the different blocks can be optimized in parallel. The problem is that even simple functions are not separable. Despite the fact that the schema theorem considers the behavior of a GA in only one step, the major lack is that the building block hypothesis has no clear formulation that can be verified or falsified.

Evolutionary Programming

Evolutionary programming (EP) (Fogel, Owens, and Walsh, 1966) considers a representation that is fit to the problem. This means that the different parameters that have to be optimized can have different codomains. The main variation operator is mutation, which can be handled very flexibly, and EP makes usually no use of crossover operators. In a standard approach, a parent population of size μ produces μ children by mutation. The new parent population consists of μ individuals from the parents and children that have been selected by a probabilistic selection method (e.g., fitness-proportional selection). In the selection step, it is important to ensure that a best individual of the parents and the children is integrated into the new parent population such that the best solution found will not get lost during the optimization process.

Genetic Programming

Genetic programming (GP) developed by Koza (1991) is an evolutionary computation approach that has become very popular in recent years. Instead of searching the considered search space, one tries to construct good computer programs that solve the given task. Therefore, individuals are possible computer programs, usually represented as trees that represent expressions. These trees are evolved during the evolution process. Similarly to the other approaches, a set of computer programs constitutes a population, and a parent population creates an offspring population using crossover and mutation. The fitness of a program is given by its performance with respect to the evaluation of some test cases. To select individuals from the parents and the children for the new parent population, often fitness-proportional selection is used.

3.2 Ant Colony Optimization

Ant colony optimization (ACO) is another bio-inspired approach to solve optimization problems. Introduced by Colorni, Dorigo, and Maniezzo (1992), it has been shown to be especially successful for solving combinatorial optimization problems. A good overview of the different techniques used in this field is given in the book of Dorigo and Stützle (2004). In contrast to EAs, where solutions are constructed from the current set of solutions, solutions are in this case obtained by random walks on a so-called construction graph which is usually a directed graph. ACO algorithms are inspired by the search of an ant colony for a common source of food. It has been noticed that ants find very quickly a shortest path to a source of food. The information about which path to take to get to the food is distributed between the ants by them leaving a piece of information, called pheromone, on the path. As longer paths to the source take much more time than shorter paths, shorter paths are more often visited. This implies larger pheromone values on shorter paths after a small amount of time.

Construction of Solutions

The above-mentioned ideas are used to solve optimization problems. Solutions of a given problem are obtained by random walks of ants on a construction graph that has positive values, the pheromone values, on the edges. These values influence the random walks in the way that edges with large values have a larger probability of being traversed. In addition, the model of ACO algorithms allows us to include heuristic information to guide the random walks. This information additionally influences the probability of which vertex to visit next in the random walk.

In an ACO algorithm, each ant of the colony exploits the construction graph to search for an optimal solution. We assume that the ant colony is a set $A = \{a_1, \ldots, a_k\}$ of k ants. Each a_i has memory that can be used to store information about the path it has followed so far. This memory can be used to build feasible solutions, compute a heuristic value η, evaluate the solution that has been found, and retrace the path backwards. An ant has a start state and one or more termination conditions. In a single step, the ant moves from a current vertex v of the construction graph to one of its successors. This move is chosen based on a probabilistic rule and depends on the pheromone values on the edges, heuristic information associated with components and connections in the neighborhood of v, the ant's private memory, and the problem constraints. When adding a component to the solution the ant builds up, the algorithm may update the pheromone value of the connection that corresponds to this solution. This is not always done. Usually the pheromone values are updated after the complete solution has been built. Here, the ant retraces the path it has taken to build the solution and increases the pheromone values along these edges.

Let $C = (V, E)$ be the construction graph of a given problem. The pheromone value of an edge $e = (u, v) \in E$ is denoted by $\tau_{(u,v)}$. In addition it is possible to assign to each edge $(u, v) \in E$ a piece of heuristic information $\eta_{(u,v)}$. We assume that an ant is at vertex u and denote the set of allowed successors by $N(u)$. Due to the problem constraints, this set may be a subset of the successors of u in C. The probability that the ant visits the vertex $v \in N(u)$ in the next step is given by

$$p_v = \frac{[\tau_{(u,v)}]^\alpha \cdot [\eta_{(u,v)}]^\beta}{\sum_{w \in N(v)} [\tau_{(u,w)}]^\alpha \cdot [\eta_{(u,w)}]^\beta}.$$

Here $\alpha, \beta \geq 0$ are parameters that determine the importance of the pheromone values and the heuristic information, respectively.

Updating Pheromone Values

In the update procedure of an ACO algorithm, the pheromone values are usually decreased by an amount that depends on the value before the update and the evaporation factor ρ, $0 \leq \rho \leq 1$. Let $\tau_{(u,v)}$ be the pheromone value on edge $(u, v) \in E$ before the update. The value is decreased to $(1 - \rho)\tau_{(u,v)}$ in a first step. This implies that information about which paths are taken so far gets lost during the run of the algorithm and helps to escape from local optima. In addition, the pheromone values on edges an ant a_i has traversed are increased by a value Δ_i that may depend on ρ as well as on the function value of the solution the ant a_i has constructed. Hence, the pheromone value $\tau'_{(u,v)}$ of edge (u, v) after the update is given by

$$\tau'_{(u,v)} = (1 - \rho)\tau_{(u,v)} + \sum_{i=1}^{k} \Delta_i.$$

There are different possibilities which ants to take into account for the update. If all ants of the colony leave pheromone values on the edges this is known as the AS-update rule. This is the update rule of the Ant System (AS) which was the first ACO algorithm proposed in the literature (Colorni et al., 1992). Using the AS-update, the amount by which an ant increases a pheromone value should depend on the function value of the constructed solution as otherwise the pheromone values are totally independent of the function f that should be optimized. If this is not the case, it would not be possible to direct the search. In the case of the IB-update rule, where IB stands for iteration-best, the ants that have constructed the best solutions of the last iteration update the pheromone values along the edges they have taken. Such an update introduces a much stronger bias towards the best solutions found so far. In the case of the best-so-far update, BS-update for short, this bias is even more extreme. Here, the pheromone values on the edges of a best solution constructed since the first iteration of the algorithm are increased in each iteration.

3.3 Other Stochastic Search Algorithms

In this section, we describe other import stochastic search algorithms that have been proposed. One important method is randomized local search (RLS). This can be seen as a simplification of the perhaps simplest evolutionary algorithms called (1+1) EA. In the case of a runtime analysis of (1+1) EA, RLS is often considered in a first step and the results are later adjusted to the EA. Local search procedures work with a predefined neighborhood and have problems if there is no better solution in this neighborhood than the current one. Then they get stuck in local optima. To escape from local optima, the Metropolis algorithm (MA) allows us to accept worsenings with a certain probability that depends on a parameter that is called the temperature. It has been shown to be useful in an approach called simulated annealing (SA) to vary this temperature over time, starting with a high temperature and cooling it down during the run of the algorithm.

3.3.1 Randomized Local Search

Apart from sampling in each iteration a search point from the given search space uniformly at random, randomized local search seems to be the simplest stochastic search algorithm that can be considered. RLS works in each iteration with one single solution s. A new solution s' is constructed from s by choosing one individual from the neighborhood of s. s is replaced by s' if s' is not inferior to s. The definition of the neighborhood is a crucial parameter. If it is too small, RLS often gets stuck in local optima. If the neighborhood is too large, even individuals that are close to the current solution may only get a too small probability of being chosen in the next step, and RLS behaves like random sampling of search points from the search space independently of s. Considering problems from the search space $\{0, 1\}^n$, RLS often uses a neighborhood that is defined by all search points that have Hamming distance 1 or 2 from the current solution s.

3.3.2 Metropolis Algorithm

In contrast to RLS, the following two approaches accept worsenings during the optimization process. The acceptance of a worsening depends on the difference between the fitness values of s and s' and on a so-called temperature T. In the case of the Metropolis algorithm (MA), this temperature is a fixed parameter and therefore constant during the optimization process. We assume that we are considering a function f that should be maximized. In the case where $f(s') \geq f(s)$ holds, s is replaced by s'. In the other case s is replaced by s' with probability $M(s, s', T) = e^{-\frac{f(s)-f(s')}{T}}$, where M is called the Metropolis function.

    MA has been subject to rigorous analysis with respect to its runtime for the NP-hard graph bisection problem (Jerrum and Sorkin, 1998). Let $G = (V, E)$

Fig. 3.1. Connected triangles with two different weight profiles

be an undirected graph where $|V|$ is even. A bisection of G is a partitioning of V into sets L and R with $|L| = |R| = n/2$. The cut width of a bisection is defined as the number of edges that have exactly one endpoint in L and one endpoint in R. One is interested in finding a bisection with minimum cut width. Jerrum and Sorkin have considered MA for finding an optimal bisection of a random graph $G = (V, E)$ where an edge between vertices of the same partition occurs with probability p and an edge between vertices of L and R occurs with probability r. In the case where $p - r = \Theta(n^{\Delta-2})$ for a parameter Δ with $3/2 < \Delta \leq 2$, such a random graph specifies with high probability a planted bisection of density r that separates L and R, which have a slightly higher density p (Bui, Chaudhuri, Leighton, and Sipser, 1984). Then it can be shown that MA for an appropriate choice of T finds the optimal solution in about $O(n^2)$ steps with high probability if $\Delta \geq 11/6$.

3.3.3 Simulated Annealing

Simulated annealing (SA) can be seen as MA that uses different temperatures during the run of the algorithm. Starting with a temperature T_0, the temperature is decreased during the optimization process according to a cooling schedule. Such a cooling schedule can be adaptive or non-adaptive. In the case of a non-adaptive cooling schedule, the temperature T_i is known in advance for all time steps i. In the case of adaptive cooling schedules, the temperature for a given time step i may depend on the history of sampled search points.

For a long time, there were only artificial example functions (Sorkin, 1991) where it could be proven that a cooling schedule can be useful in reducing the runtime significantly. Wegener (2005b) has presented the first "natural" example where this is the case. He has shown that SA can outperform MA for each fixed temperature on a class of instances of the minimum spanning tree problem. Wegener has investigated connected triangles (see Figure 3.1) with $m = 6n$ edges and $4n + 1$ vertices. The structure of this graph is the same as the triangle part of the graph we will investigate in Chapter 5 for the analysis of evolutionary algorithms until they have computed a minimum spanning tree. The number of triangles equals $2n$. Each triangle gets a weight profile (w_1, w_2, w_3), which is the ordered vector of the three edge weights. The basic idea is to construct weight profiles such that for each fixed temperature it is hard to optimize all triangles while an appropriate cooling schedule is able to optimize all triangles. Wegener uses n triangles with the weight profile $(1, 1, m)$ and n triangles with the weight profile (m^2, m^2, m^3). Then he distinguishes between high temperatures $(T \geq m)$ and low temperatures $(T < m)$. He shows that high temperatures are not able to optimize the triangles with the weight profile $(1, 1, m)$ and low temperatures are not able to optimize the triangles with weight profile (m^2, m^2, m^3) in a polynomial number of steps. Hence, different temperatures are necessary to find an optimal solution solution quickly. An optimal solution can be obtained in a polynomial number of steps by using an appropriate cooling schedule in SA.

4

Analyzing Stochastic Search Algorithms

In this chapter, we introduce the stochastic search algorithms for single-objective optimization that will be subject to the analyses throughout this book. We start by describing algorithms for single-objective optimization problems in Section 4.1. There, we consider different variants of RLS and variants of a well-known evolutionary algorithm called (1+1) EA. Afterwards, we introduce some basic methods methods for analyzing stochastic search algorithms.

4.1 Simple Stochastic Search Algorithms

In this section, we introduce the stochastic search algorithms that we will consider for single-objective optimization problems. We investigate heuristics for discrete search spaces. Most of the problems we examine in this book are graph problems where one searches for a good set of vertices or edges. In this case, solutions can be represented as binary strings where each bit corresponds to a vertex or an edge. All our algorithms are described for the minimization of a given fitness function f but can also be easily applied to problems where the goal is to maximize a given fitness function.

4.1.1 Randomized Local Search

Randomized local search (RLS) in the binary case produces from a current solution $s \in \{0,1\}^n$ a new one s' by flipping a randomly chosen bit (see Algorithm 1). We index the algorithm with the subscript "b" to indicate the binary search space and the superscript "1" to emphasize that only one bit is flipped. We will sometimes also refer to the bit-flip operators in RLS algorithms as mutation operators.

For all stochastic search algorithms, we consider no stopping criterion to be defined. In applications this is, of course, necessary. Often such an algorithm is stopped after a predefined number of iterations or if no progress has been made

F. Neumann, C. Witt, *Bioinspired Computation*
in Combinatorial Optimization, Natural Computing Series,
DOI 10.1007/978-3-642-16544-3_4, © Springer-Verlag Berlin Heidelberg 2010

Algorithm 1 RLS$_b^1$

1. Choose $s \in \{0,1\}^n$ uniform at random.
2. Choose $i \in \{1,\dots,n\}$ uniform at random and flip the ith bit of s.
3. Replace s with s' if $f(s') \leq f(s)$.
4. Repeat Steps 2 and 3 forever.

Algorithm 2 RLS$_b^{1,2}$

1. Choose $s \in \{0,1\}^n$ uniform at random.
2. Choose $b \in \{0,1\}$ uniform at random.
   If $b = 0$, choose $i \in \{1,\dots,n\}$ uniform at random and define s' by flipping the ith bit of s.
   If $b = 1$, choose $(i,j) \in \{(k,l) \mid 1 \leq k < l \leq n\}$ uniform at random and define s' by flipping the ith and the jth bits of s.
3. Replace s with s' if $f(s') \leq f(s)$.
4. Repeat Steps 2 and 3 forever.

---

for a certain number of steps. We consider the algorithms we analyze as infinite stochastic processes and are interested in the number of fitness evaluations until a given task has been achieved. In the case of exact optimization, the number of fitness evaluations until an optimal solution has been produced is investigated. Often the expectation of this value is analyzed and called the expected optimization time of the considered algorithm. Especially in the case where one cannot hope to compute optimal solutions in a polynomial number of steps, e.g., for NP-hard problems, one is interested in the number of fitness evaluations until the algorithm has produced a good approximation of an optimal solution.

Flipping one single bit is not useful for most graph problems. Often the number of 1s (or edges) is the same for all good search points, e.g., for traveling salesperson problems (TSPs) or minimum spanning trees. Then, all Hamming neighbors of good search points are bad, implying that we have many local optima. Therefore, we work with the larger neighborhood of Hamming distance 2 and investigate a variant of randomized local search given in Algorithm 2. This time the superscript "1,2" is used for clarification.

4.1.2 A Simple Evolutionary Algorithm

The evolutionary algorithms that we consider for single-objective optimization problems use a population of size 1 and produce at each time step one single child. They can be seen as variants of RLS, which we introduced in the last section, with a more flexible mutation operator. Usually, a mutation operator in this scenario should be able to search globally. Here, each search point of the considered search space should get a positive probability of being chosen

---

Algorithm 3 $(1+1)$ EA_b

---

1. Choose $s \in \{0,1\}^n$ uniform at random.
2. Produce s' by flipping each bit of s independently of the other bits with probability $1/n$.
3. Replace s with s' if $f(s') \leq f(s)$.
4. Repeat Steps 2 and 3 forever.

---

in the next step. Again we consider the algorithm for the search space $\{0,1\}^n$ first. The perhaps simplest evolutionary algorithm that can be considered in this case is $(1+1)$ EA_b. Starting with a randomly chosen bitstring s of length n, the algorithm produces in each iteration a child by flipping each bit of s with probability $1/n$. $(1+1)$ EA_b for minimizing a fitness function f is given in Algorithm 3.

$(1+1)$ EA_b has been the subject of the first analyses of evolutionary algorithms with respect to their expected optimization time. In the beginning, the behavior of this algorithm on pseudo-boolean functions that depend on n variables was considered. Some of first main results were obtained by Droste et al. (2002). It has been shown that the expected time to reach an optimal search point by this algorithm in the considered search space is always bounded above by n^n, as the probability to choose an optimal search point in the next step is at least n^{-n}. More detailed analyses consider pseudo-boolean functions with different properties. One major result is that the expected optimization time on linear functions is $O(n \log n)$. The class of functions of degree 2 is too huge to get a polynomial upper bound on the runtime for each function, as optimizing polynomials of degree at least 2 is *NP*-hard.

4.1.3 Algorithms for Multi-Objective Optimization

In this case of multi-objective optimization, one searches for a set of optimal solutions instead of a single one. We want to examine multi-objective evolutionary algorithms (MOEAs) that are generalizations of RLS_b^1 and $(1+1)$ EA_b. Therefore, we investigate and analyze a simple algorithm called SEMO (Simple Evolutionary Multi-Objective Optimizer) due to Laumanns, Thiele, and Zitzler (2004).

The fitness of a search point s is given by a vector $f(s) = (f_1(s), \ldots, f_k(s))$. W. l. o. g., we assume that each function f_i should be minimized and write $f(s) \leq f(s')$ iff $f_i(s) \leq f_i(s')$ holds for all i, $1 \leq i \leq k$. A solution s domintates a solution s' iff $f(s) \leq f(s')$ and $f(s) \neq f(s')$ holds. If s dominates s' we also say that $f(s)$ dominates $f(s')$.

SEMO (see Algorithm 4) starts with an initial solution $s \in \{0,1\}^n$ that is chosen uniformly at random. All non-dominated solutions are stored in the population P. In each step a search point from P is chosen uniformly at random and one bit is flipped to obtain a new search point s'. The new

---

Algorithm 4 SEMO

---

1. Choose an initial solution $s \in \{0, 1\}^n$ uniformly at random.
2. Determine $f(s)$ and initialize $P := \{s\}$.
3. Repeat
    a) Choose $s \in P$ randomly.
    b) Choose $i \in \{1, \ldots, n\}$ randomly.
    c) Define s' by flipping the ith bit of s.
    d) Determine $f(s')$,
    e) Let P be unchanged, if there is an $s'' \in P$ such that $f(s'') \leq f(s')$ and $f(s'') \neq f(s')$
    f) Otherwise, exclude all s'' where $f(s') \leq f(s'')$ from P and add s' to P.

Algorithm 5 Global SEMO (GSEMO)

---

1. Choose an initial solution $s \in \{0, 1\}^n$ uniformly at random.
2. Determine $f(s)$ and initialize $P := \{s\}$.
3. Repeat
    a) Choose $s \in P$ randomly.
    b) Define s' by flipping each bit of s independently of the other bits with probability $1/n$.
    c) Determine $f(s')$,
    d) Let P be unchanged, if there is an $s'' \in P$ such that $f(s'') \leq f(s')$ and $f(s'') \neq f(s')$
    e) Otherwise, exclude all s'' where $f(s') \leq f(s'')$ from P and add s' to P.

---

population contains for each non-dominated fitness vector $f(s)$, $s \in P \cup \{s'\}$, one corresponding search point, and in the case where $f(s')$ is not dominated s' is chosen.

Applying SEMO to a single-objective optimization problem, we obtain RLS_b^1 where in each step a single bit is flipped. Giel (2003) has introduced an algorithm called Global SEMO (GSEMO), which is shown in Algorithm 5. This algorithm differs from SEMO by using the more general mutation operator of $(1+1)$ EA$_b$. GSEMO applied to single-objective optimization problems equals $(1+1)$ EA$_b$.

We will analyze the algorithms until they have achieved certain goals for different combinatorial optimization problems. In the case of polynomially solvable problems, we are interested in the time until a solution is produced for each Pareto optimal objective vector, whereas in the case of *NP*-hard problems we are interested in the time to achieve a good approximation of the Pareto front. We will also examine how multi-objective models of single-objective optimization problems can help to speed up stochastic search algorithms. In this case, we are mainly interested in the quality of a particular solution in the population, namely the one solving the single-objective optimization problem.

4.2 Basic Methods for the Analysis

Until the early 1990s, theory on evolutionary algorithms mainly dealt with the convergence of EAs or results that showed the behavior of an EA in one single iteration. The first runtime analysis of an EA was given by Mühlenbein (1992). Evolutionary algorithms are stochastic search algorithms, but for a long time they were not analyzed in the way randomized algorithms normally are. The main reason for this is that the people who worked on theoretical aspects of evolutionary computation had a different background than people in theoretical computer science or discrete mathematics. With regard to evolutionary algorithms as a class of randomized algorithms, a lot of strong methods are available. Such methods have already been applied in the field of randomized algorithms (Motwani and Raghavan, 1995). A very important issue when analyzing the runtime of EAs is the application of large deviation inequalities such as Chernoff bounds or Markov's inequality. Another useful method is to follow the considerations of the coupon collectors problem. Since the mid-1990s, a lot of new methods for analyzing the runtime of EAs have been obtained. In this section, we want to discuss some important methods that have been used. These methods will be applied in our analysis of evolutionary algorithms for combinatorial optimization problems.

To show how to apply different methods that have been developed, we consider the class of linear pseudo-boolean functions. A linear pseudo-boolean function $f\colon \{0,1\}^n \to \mathbb{R}$ is defined by

$$f(x) \;=\; w_1 x_1 + w_2 x_2 + \cdots + w_n x_n,$$

where $w_i \in \mathbb{Z}$.

W. l. o. g., we assume that all w_i attain non-negative values. The case of (partially) negative weights can be handled analogously to the following investigations, as a weight $w_i \neq 0$ determines independently of the other weights whether the bit x_i has to be set to 1 or 0 in an optimal solution. In the case where some weights are 0, the function value does not depend on the corresponding bits. The upper bound given in Theorem 4.4 also holds in this case, but the lower bound given in Theorem 4.3 needs the condition that there are $\Theta(n)$ weights distinct from 0.

4.2.1 Fitness-Based Partitions

This simple method has been used for a wide class of problems. We assume that we are considering a stochastic search algorithm that works in each iteration with one solution that produces one offspring. All variants of RLS and (1+1) EA we have discussed in Section 4.1 fit into this scenario. Assume that we are working in a search space S and consider w. l. o. g. a function $f\colon S \to \mathbb{R}$ that should be maximized. S is partitioned into disjoint sets $A_1, \ldots, A_m$ such that $A_1 <_f A_2 <_f \cdots <_f A_m$ holds, where $A_i <_f A_j$ means that $f(a) < f(b)$

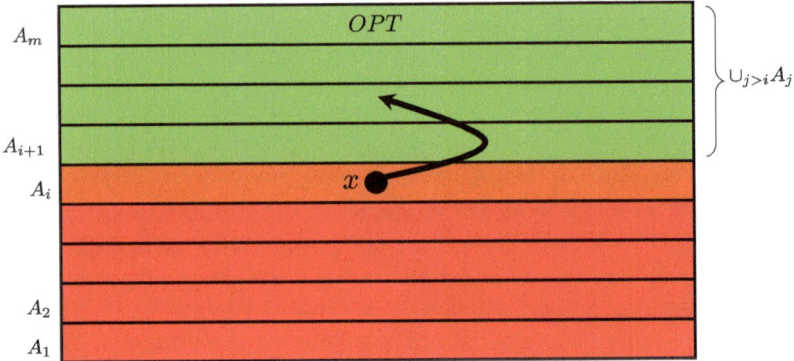

Fig. 4.1. Illustration of fitness-based partitions

holds for all $a \in A_i$ and all $b \in A_j$. In addition, A_m contains only optimal search points. An illustration is given in Figure 4.1. We denote for a search point $x \in A_i$ by $p(x)$ the probability that in the next step a solution $x' \in A_{i+1} \cup \cdots \cup A_m$ is produced. Let $p_i = \min_{a \in A_i} p(x)$ be the smallest probability of producing a solution with a higher partition number.

Lemma 4.1. *The expected optimization time of a stochastic search algorithm that works at each time step with a population of size 1 and produces at each time step a new solution from the current solution is upper bounded by $\sum_{i=1}^{m-1}(1/p_i)$.*

Proof. The expected time of a success for independent Bernoulli trials with probability p is $1/p$. Hence, the expected time to produce from a search point $x \in A_i$ a search point x' with $x' \in A_j$, $j > i$, is upper bounded by $1/p_i$. This implies that the expected time until an optimal search point has been produced is upper bounded by $\sum_{i=1}^{m-1}(1/p_i)$. $\square$

To come up with good upper bounds using this method, one has to use a good partitioning of the search space such that there are not too many partitions and that there is a high probability of leaving the current partition and producing a search point in a better one.

We consider a simple example. OneMax: $\{0,1\}^n \rightarrow \mathbb{R}$ is a simple linear pseudo-boolean function where $w_i = 1$, $1 \le i \le n$, holds. It is defined by OneMax$(x) = \sum_{i=1}^n x_i$ and should be maximized. The function returns for a bitstring x of length n the number of 1s in x. We consider (1+1) EA$_b$ for maximization problems, where a new solution is accepted if its fitness value is not smaller than the value of the best solution up to now.

Theorem 4.2. *The expected optimization time of (1+1) EA$_b$ on OneMax is $O(n \log n)$.*

Proof. The search space is partitioned into $n + 1$ sets $A_0, \ldots, A_n$ where A_i contains all solutions x with $\mathrm{OneMax}(x) = \mathrm{i}$. Assume that the currently best solution x belongs to A_{n-k}. Then there are exactly k 0-bits that can be flipped to obtain an improvement. The probability of an improvement in the next step is at least $\frac{k}{n}(1 - \frac{1}{n})^{n-1} \geq \frac{k}{en}$. Hence, the expected waiting time for an improvement is upper bounded by en/k. Summing up the waiting times for the different values of k we get

$$\sum_{k=1}^{n} \frac{en}{k} = en \cdot \sum_{k=1}^{n} \frac{1}{k} = O(n \log n). \qquad \square$$

In the case where one works with a larger population, often an individual with the highest partition number in the population is considered. Then one can analyze the time until this individual has become an optimal one. The method works in nearly the same way as in the case of a population of size 1, but one often has to add an additional factor to choose the right individual in the next step.

4.2.2 Chernoff Bounds and Coupon Collectors

Large deviation inequalities have widely been used in the analysis of randomized algorithms. In the case of stochastic search algorithms, they are often useful for showing the typical behavior of such a heuristic. We consider (1+1) $\mathrm{EA_b}$ which chooses the initial solution x uniformly at random from $\{0, 1\}^n$ by setting each bit with equal probability to 0 or 1. Hence, n Bernoulli trials are considered where $\mathrm{Prob}(x_i = 1) = \mathrm{Prob}(x_i = 0) = 1/2, 1 \leq i \leq n$, holds. The expected number of 1s in the initial solution is therefore $n/2$ and there are at most $2n/3$ 1s in the initial bitstring with probability $1 - e^{-\Omega(n)}$ using Chernoff bounds (see Appendix A.5).

In the coupon collector's problem (Motwani and Raghavan, 1995), n different coupons are given and at each time step a coupon is chosen uniformly at random from among all coupons. Let t be the number of trials. Then one studies the number of trials until each of the n coupons has been chosen at least once. The expected number of trials until each coupon has been chosen at least once is $\Theta(n \log n)$ (see Appendix A.13). Using Chernoff bounds and the ideas of the coupon collectors problem, it is easy to obtain a lower bound of $\Omega(n \log n)$ on each linear pseudo-boolean function with non-zero weights. To show how to use Chernoff bounds and the ideas of the coupon collectors problem, we present the proof which can be found in Droste et al. (2002). W.l.o.g , we assume that all weights attain positive values. Hence, the only optimal solution is the bitstring $(1, \ldots, 1)$.

Theorem 4.3. *The expected optimization time of (1+1) EA_b on each linear pseudo-boolean function with non-zero weights is $\Omega(n \log n)$.*

Proof. Using Chernoff bounds, the expected number of 0s in the initial bit-string is at least $n/3$ with probability $1 - e^{-\Omega(n)}$. To obtain the proposed lower bound, we analyze the expected time until each of the 0-bits has been flipped at least once under the condition that there are at least $n/3$ 0-bits after initialization. This is done in a similar fashion as in the case of the coupon collector's theorem.

Let t be a specific number of steps. The probability that a specific 0-bit has not been flipped at least once in t steps is $(1 - 1/n)^t$. Hence, the probability that it has flipped at least once in t steps is $1 - (1 - 1/n)^t$, and the probability that each of the $n/3$ 0-bits has flipped at least once is $(1 - (1 - 1/n)^t)^{n/3}$. The probability that at least one of the $n/3$ 0-bits has never flipped during t steps is $1 - (1 - (1 - 1/n)^t)^{n/3}$. Hence, the probability that at least one 0-bit has not been flipped during $t = (n - 1)\ln n$ steps is $1 - (1 - (1 - 1/n)^{(n-1)\ln n})^{n/3} \geq 1 - e^{-1/3}$.

Altogether, the optimization time of (1+1) EA$_b$ is $\Omega(n \log n)$ with probability at least $1 - e^{-1/3} - e^{-\Omega(n)} = \Omega(1)$, which proves the theorem. $\square$

4.2.3 Expected Multiplicative Distance Decrease

The method of the expected multiplicative distance decrease has been developed to analyze the runtime behavior of stochastic search algorithms on problems with a large number of different values that the fitness function may attain. For example, this is the case for the minimum spanning tree problem where an exponential number of spanning trees with different weights is possible.

The method is illustrated in Figure 4.2. It can be applied to problems where we are able to transform each solution s into an optimal solution s_{opt} by a set $O = \{o_1, \dots, o_r\}$ consisting of r operations that all have the same probability of happening in the next step. We assume that this probability can be lower bounded by α and that the set of possible fitness values contains only integers. For simplicity, the value of r does not change in all considerations.

Note that the number of operations until s_{opt} has been reached depends on the solution s. W.l.o.g., we assume that $O' = \{o_1, \dots, o_{r_1}\}$, $O' \subseteq O$, is the set of operations necessary to turn s into s_{opt}. Then $r - r_1$ operations are added such that one can work at each time step with the same value of r. It is important that the application of each of the operations of O' lead to a solution s' that is not inferior to s. This implies that each operation of O' applied to s is accepted. W.l.o.g., we assume that the considered fitness function f should be maximized. Let $d = f(s_{\text{opt}}) - f(s)$ be the distance (measured as the difference of the function values) of s from an optimal one. As all operations have the same probability, the expected decrease in the distance when producing a solution s' by an operation that is chosen uniformly at random from the set O is at least $\frac{f(s_{\text{opt}}) - f(s)}{r}$. Note that non-accepted operations of $O \setminus O'$ contribute a distance decrease of 0. The expected distance of s' from s_{opt} is

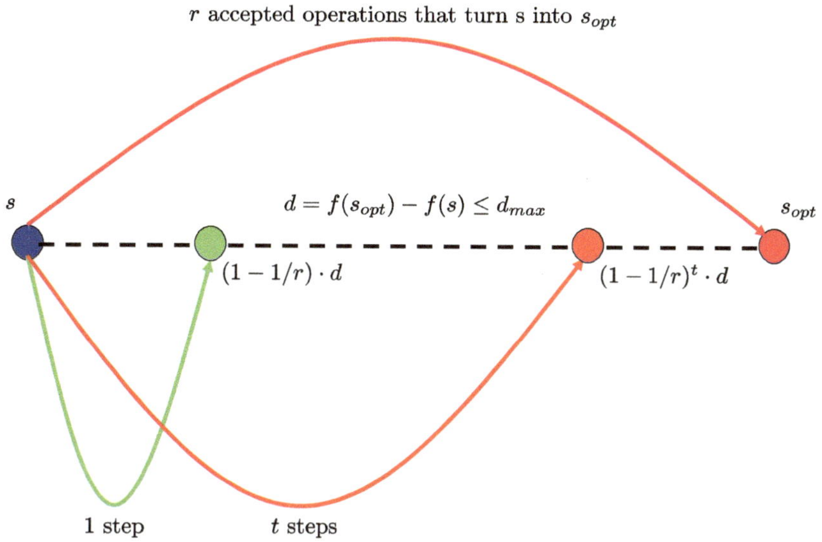

Fig. 4.2. Illustration of the expected multiplicative distance decrease

$$(1 - 1/r) \cdot (f(s_{\mathrm{opt}}) - f(s))$$

after 1 step, and the expected distance after t such steps is

$$(1 - 1/r)^t \cdot (f(s_{\mathrm{opt}}) - f(s)).$$

Let

$$d_{\mathrm{max}} = \max_{s \in \{0,1\}^n} (f(s_{\mathrm{opt}}) - f(s))$$

be the maximum distance of any search point in the search space from an optimal one. After having executed t randomly chosen operations of O, the expected distance to an optimal solution is at most $(1 - 1/r)^t \cdot d_{\mathrm{max}}$. Choosing $t = c \cdot r \cdot \log d_{\mathrm{max}}$, c an appropriate constant, the expected distance is at most $1/2$. Using Markov's inequality (see Appendix A.4), the probability that the distance is at least 1 is upper bounded by $1/2$. As the set of possible fitness values contains only integers, the probability of having achieved an optimal solution (i.e., the distance is 0) is at least $1/2$. This implies that the expected number of operations belonging to the set O until an optimal solution has been achieved is at most $2t = O(r \cdot \log d_{\mathrm{max}})$. The probability of an operation belonging to the set O is at least $r \cdot \alpha$. Using this, the expected optimization time is $O((r \cdot \alpha)^{-1} r \cdot \log d_{\mathrm{max}}) = O(\alpha \cdot \log d_{\mathrm{max}})$.

We consider linear pseudo-boolean functions and define $w_{\mathrm{max}} = \max_i |w_i|$. Applying the method of expected multiplicative distance decrease, we show in a simple way an upper bound on the expected optimization time of $(1+1)\,\mathrm{EA}_b$ on each linear pseudo-boolean function, which is according to Theorem 4.3

optimal as long as the weights are polynomially bounded in n. W.l.o.g., we assume that $w_i \geq 0$, $1 \leq i \leq n$, holds.

Theorem 4.4. *The expected optimization time of (1+1) EA_b on linear functions is upper bounded by $O(n(\log n + \log w_{\max}))$.*

Proof. The set of operations O contains all steps where only one single bit flips. Hence, O contains $r = n$ operations. The set O' contains all operations flipping one single 0-bit. As $w_i \geq 0$, $1 \leq i \leq n$, each operation of O' is accepted. The probability of one specific operation of O is $1/n \cdot (1 - 1/n)^{n-1} \geq 1/(en) := \alpha$ and $d_{\max} \leq n \cdot w_{\max}$ holds. Using the method of expected multiplicative distance decrease, the expected optimization time is upper bounded by $O(n \log d_{\max}) = O(n(\log n + \log w_{\max}))$. $\square$

Note that the given upper bound is $O(n \log n)$ as long as all weights are polynomially bounded in n. It is possible to obtain a more general upper bound of $O(n \log n)$ even if the weights are not polynomially bounded. This proof is much more complicated than the one presented here and can be found in Droste et al. (2002).

4.2.4 Cover Time of Random Walks

In the following, we show how classical results on random walks on a given graph can be used for the analysis of stochastic search algorithms. In particular, we show how results on the cover time of a random walk can be directly used to give bounds on the runtime of this class of algorithms when dealing with plateau functions.

Plateaus are regions in the search space where all search points have the same objective vectors. Consider a function $f \colon \{0,1\}^n \to \mathbb{R}$ and assume that the number of different objective values for that function is N. Then there are at least $2^n/N$ search points with the same objective value. Often, the number of different objective values for a given function is polynomially bounded. This implies an exponential number of solutions with the same objective value. Nevertheless, such functions where N is polynomially bounded are easy to optimize for evolutionary algorithms if for each non-optimal solution there is a better Hamming neighbor, which means that an improvement can be made by flipping a single bit of a non-optimal solution. Polynomial upper bounds for such functions and typical stochastic search algorithms can be obtained by using the method of fitness-based partitions introduced in Section 4.2.1.

If this is not the case, the search for a stochastic search algorithm may become much harder. In the extreme case, we end up with the function *NEEDLE*, where only one single solution has objective value 1 and the remaining ones get an objective value of 0. Here, typical stochastic search algorithms require an exponential number of steps to reach the optimal solution as the function does not give any hints towards the optimum. The behavior of (1+1) EA_b

---

Algorithm 6 Random Walk on a graph $G = (V, E)$

Start at a vertex $v \in V$.
repeat
    Choose a neighbor w of v in G uniformly at random.
    Set $v := w$.
until stop

---

on plateaus of different structures has been studied by Jansen and Wegener (2001).

We want to relate the behavior of stochastic search algorithms on plateau functions to random walks on a given graph, and we consider the following problem. Given a connected graph $G = (V, E)$, a random walk starts at a vertex $v \in V$ and moves in each step to a neighbor of the current vertex that is chosen uniformly at random from among all neighbors. An algorithm describing this random walk procedure is stated in Algorithm 6.

Definition 4.5. *Given an undirected connected graph $G = (V, E)$, the cover time of a random walk on G is the number of steps until each vertex $v \in V$ has been visited at least once.*

The following result has been obtained by Aleliunas, Karp, Lipton, Lovász, and Rackoff (1979).

Theorem 4.6 (Upper bound for Cover Time). *Given an undirected connected graph $G = (V, E)$ with n vertices and m edges, the cover time is upper bounded by $2|E|(|V| - 1)$.*

To illustrate how to use this bound for the analysis of stochastic search algorithms, we consider the function SPC (short path with constant values), which was introduced by Jansen and Wegener (2001). Let $|x|_0$ denote the number of zeros in a bitstring x. The function SPC is defined as

$$SPC(x) := \begin{cases} |x|_0 & : \quad x \notin \{1^i 0^{n-i}, 0 \leq i \leq n\} \\ n+1 & : \quad x \in \{1^i 0^{n-i}, 0 \leq i < n\} \\ 2n & : \quad x = 1^n. \end{cases}$$

We denote by $SP := \{1^i 0^{n-i}, 0 \leq i < n\}$ the set of search points that constitute the plateau of fitness $n + 1$. Consider the graph $G_{SP} = (V, E)$ with $V = \{v_0, v_1, \ldots, v_n\}$ shown in Figure 4.3. The vertex v_i corresponds to the search point $1^i 0^{n-i}$. The edge set is given by $E = \{\{v_i, v_{i+1}\}, 0 \leq i \leq n-1\}$, i.e., an edge is present between two vertices if the corresponding search points have Hamming distance 1. The graph G_{SP} consists of $n + 1$ vertices and n edges. Hence, the cover time of a random walk on G_{SP} is upper bounded by $2n^2$. This observation is very useful to bound the expected optimization time of RLS_b^1 on SPC as shown in the following theorem.

Fig. 4.3. Graph G_{SP} and corresponding search points of $SP \cup \{1^n\}$

We want to point out the relation between the search process of RLS_b^1 on $SP \cup \{1^n\}$ and the random walk on G_{SP}. We identify search points with their corresponding vertices in the graph G_{SP}. Note that once a solution with a corresponding vertex in this graph has been obtained, no search point that does not have a corresponding vertex in the graph is accepted. The optimum has been reached if the vertex v_n has been visited for the first time. We call a step of the algorithm relevant if it is accepted by the algorithm. RLS_b^1 always flips exactly one bit in each mutation step. Hence, a relevant step consists of moving to a neighbor of the current solution in the graph. This step is unique in the case where the current solution is 0^n. Then, only the mutation step flipping the first bit is accepted. For a search point x corresponding to a vertex v_i, $1 \le i \le n-1$, the probability of moving to v_{i-1} as well as the probability of moving to v_{i+1} is $1/n$, as the bit x_i or the bit x_{i+1} has to be flipped. Both accepted mutation steps occur with the same probability. Hence, in the next mutation step, the neighbor of v_i is chosen uniformly at random.

In summary, we have shown that with regard to the relevant steps when we are at a vertex v_i, $1 \le i \le n-1$, RLS_b^1 acts like the random walk algorithm on the graph G_{SP}. This implies that the expected number of relevant steps until the solution 1^n has been obtained for the first time is at most $2n^2$ after a solution of $SP \cup \{1^n\}$ has been obtained for the first time.

Theorem 4.7. *The expected optimization time of RLS_b^1 on SPC is upper bounded by $O(n^3)$.*

Proof. As long as no search point of $SP \cup \{1^n\}$ has been produced, RLS_b^1 maximizes the number of zeros in the bitstring. RLS_b^1 behaves as on the function OneMax. The only difference is that it maximizes the number of zeros instead of the number of 1s. Hence, after an expected number $O(n \log n)$ steps, a solution of $SP \cup \{1^n\}$ is obtained using similar arguments as those in the proof for (1+1) EA on OneMax (see Theorem 4.2).

We already know that the expected number of relevant steps to reach the optimum after having reached a solution of $SP \cup \{1^n\}$ is upper bounded by $2n^2$. A relevant step happens with probability at least $1/n$ in the next mutation step, and the expected waiting time for such a step is therefore upper bounded by n. Hence, after an expected number of at most $2n^3$ steps, the optimum is found after a search point of $SP \cup \{1^n\}$ is first produced. This completes the proof. □

4.2.5 Gambler's Ruin Theorem and Drift Analysis

Closely related to the previous discussion of random walks is the analysis of a simple combinatorial game, whose basic properties often reappear in the stochastic processes induced by search algorithms. This time the random walk is not necessarily "fair," i.e., not all neighboring states are necessarily chosen with the same probability. Typically, the game is formulated as a Markov process on the state space $S = \{0, \dots, b\}$, where 0 and b are absorbing states. In state i, $1 \leq i \leq b - 1$, the probability of moving to state $i + 1$ is denoted by p, and the probability of going to state $i - 1$ by $q := 1 - p$. Hence, the process necessarily changes state by $+1$ or -1 until an absorbing state has been reached, and the transition probabilities are the same for all non-absorbing states. Starting in state a, $1 \leq a \leq b - 1$, we are interested in the absorbing state that is eventually reached. Interpreting the state space as the capital of a gambler and b as the capital of the bank, we are confronted with a game where the gambler either wins or loses one unit of money with a certain probability in each step until either the bank or he is ruined.

The following list of results is commonly subsumed under the headline "gambler's ruin theorem" (Feller, 1968). We only state those propositions that will be relevant in the course of this book.

Theorem 4.8 (Gambler's Ruin Theorem). *If $p = q = 1/2$, the probability of the gambler's game ending at state 0 equals*

$$q_a = 1 - \frac{a}{b},$$

and the expected duration of the game is

$$D_a = a(b - a).$$

If $q \neq p$ then

$$q_a = \frac{r^b - r^a}{r^b - 1},$$

and

$$D_a = \frac{a}{q - p} - \frac{b}{q - p} \cdot \frac{r^a - 1}{r^b - 1}$$

where $r = q/p$.

The theorem is often used in the case $q > p$, i.e., when there is a tendency towards decreasing the state. We see that q_a becomes close to 1 in this case, particularly if $b \gg a$. If the initial capital is low compared to the capital of the bank and the game is in favor of the bank, the probability of the gambler's ruin is high. This scenario reappears in situations where stochastic search algorithms tend to walk towards an undesired state. Due to the nature of the above expression, the gambler's ruin theorem allows then for exponential lower bounds (with respect to the parameter b) on the optimization time.

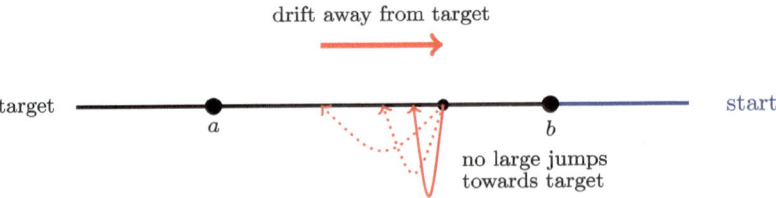

Fig. 4.4. Illustration of the scenario underlying the drift theorems for lower bounds

A drawback of the gambler's ruin theorem is that it restricts the change of state to either $+1$ or -1, i.e., assumes a local behavior of the process. Therefore, it can often be directly applied to processes induced by RLS_b^1 while it is not well suited to model the behavior of the $(1+1)$ EA, which is allowed to flip all bits in a step. This allows us, in principle, to move from any state to any other state in a single step. Still, the $(1+1)$ EA is inclined to perform only small changes. Therefore the intuition of the gambler's ruin theorem can still be carried over in many cases. Since it is more convenient to deal with a positive value for the expected direction of the movement, we now turn things around and wait for a Markov process to reach the lower limit of the interval $[a, b]$ given a starting point above state b. With the aim of showing that the whole interval is not passed in exponential time, we intuitively need the following two conditions (see also Figure 4.4):

- In the interval at time t, there must be a drift, an expected displacement, towards increasing the state. This will be made precise by the first condition of the following theorem.
- Drift alone is not enough. In exponentially long phases, the probability must be exponentially small of leaving the interval towards the optimum using large jumps. The random step length towards the optimum has to exhibit exponential decay, which is formalized by the second condition.

The idea behind the following theorem goes back to Hajek (1982). The variant presented here is due to Oliveto and Witt (2008).

Theorem 4.9 (Simplified Drift Theorem). *Let X_t, $t \geq 0$, be the random variables describing a Markov process over a finite state space $S \subseteq [0, N]$ and denote $\Delta_t(i) := (X_{t+1} - X_t \mid X_t = i)$ for $i \in S$ and $t \geq 0$. Suppose there exist an interval $[a, b]$ in the state space, two constants $\delta, \epsilon > 0$ and, possibly depending on $\ell := b - a$, a function $r(\ell)$ satisfying $1 \leq r(\ell) = o(\ell/\log(\ell))$ such that for all $t \geq 0$ the following two conditions hold:*

1. $E(\Delta_t(i)) \geq \epsilon$ *for $a < i < b$,*

2. $\mathrm{Prob}(\Delta_t(i) \leq -j) \leq \frac{r(\ell)}{(1+\delta)^j}$ *for $i > a$ and $j \in \mathbb{N}_0$.*

Then there is a constant $c^* > 0$ such that for $T^* := \min\{t \geq 0 \colon X_t \leq a \mid X_0 \geq b\}$ it holds that $\mathrm{Prob}(T^* \leq 2^{c^* \ell/r(\ell)}) = 2^{-\Omega(\ell/r(\ell))}$.

The theorem contains a sharp concentration result for the random variable T^*. Not only is the expected first hitting time for states less than a (given starting state at least b) exponential, but the exponential time holds also with high probability.

As an example of an application of the simplified drift theorem, reconsider the function

$$NEEDLE(x) = \begin{cases} 1 & \text{if } x = 1^n, \\ 0 & \text{otherwise.} \end{cases}$$

mentioned in Section 4.2.4.

Informally speaking, the $NEEDLE$ function is difficult since every search point except the all-1s string has the same value. Therefore, the simple search algorithms (1+1) EA_b and RLS_b^1 walk randomly on a plateau of exponential size and tend to sample search points that have only about half the bits correct for a long time. It is well known that the algorithms need expected optimization time $2^{\Omega(n)}$ (Garnier, Kallel, and Schoenauer, 1999). Using the simplified drift theorem, we can give a short proof for this result.

Theorem 4.10. *The optimization time of RLS_b^1 and (1+1) EA_b on the NEEDLE function is at least $2^{\Omega(n)}$ with probability $1 - 2^{-\Omega(n)}$.*

Proof. We set $a := 0$, $b := n/3$ and denote by X_t, $t \geq 0$, the number of zero-bits in the search point at time t. By Chernoff bounds, the initial value X_0 satisfies $X_0 \geq b$ with probability $1 - 2^{-\Omega(n)}$.

Let us consider some X_t such that $X_t = i$ for $a < i < b$. Both algorithms flip each bit (not necessarily independently) with probability $1/n$. Using the linearity of expectation, the expected number of 0-bits flipped equals i/n and the expected number of 1-bits flipped is $(n - i)/n$. We obtain

$$E(\Delta_t(i)) = \frac{n-i}{n} - \frac{i}{n} = \frac{n-2i}{n} \geq \frac{1}{3},$$

where we used $i < b$. This already establishes the first condition of Theorem 4.9 for both search algorithms.

The second condition is almost trivial to prove for RLS_b^1 since its mutation operator guarantees that $\text{Prob}(\Delta_t(i) < -1) = 0$. We set $r(\ell) = 2$ and $\delta = 1$, which implies that $\frac{r(\ell)}{(1+\delta)^j} \geq 1$ for $j \in \{0,1\}$. Hence, $\text{Prob}(\Delta_t(i) = -j) \leq \frac{r(\ell)}{(1+\delta)^j}$ for $j \in \mathbb{N}_0$. For (1+1) EA_b, we observe that the probability of flipping at least j bits in a single step is at most

$$\binom{n}{j}\left(\frac{1}{n}\right)^j \leq \frac{n^j}{j!}\left(\frac{1}{n}\right)^j = \frac{1}{j!} \leq \frac{2}{2^j}.$$

Thus, the above choices for $r(\ell)$ and δ also work for (1+1) EA_b. We have established the second condition for both algorithms.

Since optimizing $NEEDLE$ is equivalent to reaching an X_t-value of 0, Theorem 4.9 and the assumption $X_0 \geq b$ together yield that the optimization time of the search algorithms is at least $2^{\Omega(b-a)} = 2^{\Omega(n)}$ with probability at least $(1 - 2^{-\Omega(n)}) \cdot (1 - 2^{-\Omega(b-a)}) = 1 - 2^{-\Omega(n)}$. $\square$

Conclusions

We have defined basic evolutionary algorithms for single and multi-objective optimization that are often used in their complexity analysis. These algorithms are simplified algorithms but capture important features of algorithms that are used in practice and allow us to treat them in a rigorous fashion. Furthermore, we have introduced basic tools for the analysis of stochastic search algorithms and exemplified these methods by presenting results on some pseudo-Boolean functions. We will make use of the methods presented in this chapter during our investigations of combinatorial optimization problems that are carried out in the remaining part of this book.

Part II

Single-objective Optimization

5

Minimum Spanning Trees

In this chapter, we study the behavior of stochastic search algorithms on an important graph problem. We consider the well-known problem of computing a minimum spanning tree in a given undirected connected graph with n vertices and m edges. The problem has many applications in the area of network design. Assume that we have n computers that should be connected with minimum cost, where costs of a certain amount occur when one computer is connected to another one. The cost for a connection can, for example, be the distance between two considered computers. One needs to make $n-1$ connections between these computers such that all computers are able to communicate with each other. Considering a graph as a model for a possible computer network, it has n vertices and one searches for the set of edges with minimal cost that makes the graph connected.

This classical minimum spanning tree (MST) problem has the following description. Given an undirected connected graph $G = (V, E)$ on n vertices and m weighted edges, find an edge set $E' \subseteq E$ of minimal weight, that connects all vertices. The weight of an edge set is the sum of the weights of the considered edges. Weights are positive integers. Therefore, the solution is a tree on V, a so-called spanning tree. One can also consider graphs which are not necessarily connected. Then the aim is to find a minimum spanning forest, i.e., a collection of spanning trees on the connected components. All our results hold also in this case. To simplify the presentation, we assume that G is connected.

The famous algorithms due to Kruskal (1956) and Prim (1957) have worst-case runtimes of magnitude $O((n+m)\log n)$ and $O(n\log n + m)$, respectively; see any textbook on efficient algorithms (Cormen et al., 2001; Mehlhorn and Sanders, 2008). Karger, Klein, and Tarjan (1995) have given a randomized greedy algorithm that computes a minimum spanning tree in time $O(m)$ with high probability. Greedy algorithms use global ideas. Considering only the neighborhoods of two vertices u and v, it is not possible to decide whether the edge $\{u, v\}$ belongs to some minimum spanning tree. Therefore, it is interesting to analyze the runtimes obtainable by more or less local search heuristics

F. Neumann, C. Witt, *Bioinspired Computation*
in Combinatorial Optimization, Natural Computing Series,
DOI 10.1007/978-3-642-16544-3_5, © Springer-Verlag Berlin Heidelberg 2010

like randomized local search and evolutionary algorithms. We present such results, due to Neumann and Wegener (2007).

One goal is to estimate the expected time until a better spanning tree has been found. For large weights, there may be exponentially many spanning trees with different weights, which means that the distance from a starting solution to an optimal one may be exponentially. Then it is important to know how much progress a stochastic search algorithm can make with respect to an optimal solution. Therefore, we have to analyze how much better the better spanning tree is. To do this, we make use of the method of expected multiplicative distance decrease. Infact this method has been developed for analyzing stochastic search algorithms until they have computed a minimum spanning tree. In Chapter 7, we will see that this method can also be used for analyzing stochastic search algorithms on an *NP*-hard scheduling problem.

Having analyzed evolutionary algorithms for the minimum spanning tree problem, we turn to ant colony optimization. It is widely assumed and observed in experiments that the choice of the construction graph has a great effect on the runtime behavior of an ACO algorithm. The first runtime analyses of ACO algorithms for the optimization of pseudo-Boolean functions were carried out in Doerr, Neumann, Sudholt, and Witt (2007c); Gutjahr (2007); Neumann, Sudholt, and Witt (2009); Neumann and Witt (2009). The construction graph used in these papers is a general one for the optimization of pseudo-Boolean functions and does not take knowledge about the given problem into account. ACO algorithms have the advantage that more knowledge about the structure of a given problem can be incorporated into the construction of solutions. This is done by choosing an appropriate construction graph together with a procedure which allows us to obtain feasible solutions. The choice of such a construction graph together with its procedure has been observed experimentally as a crucial point for the success of such an algorithm.

We examine ACO algorithms that work on construction graphs which seem to be more suitable for the MST problem. The results we present are due to Neumann and Witt (2010). First, we consider the input graph itself. It is well known how to choose a spanning tree of a given graph uniformly at random by using random walk algorithms (Broder, 1989; Wilson, 1996). Our construction procedure produces solutions by a variant of Broder's algorithm. We show a polynomial, but relatively large, upper bound for obtaining a minimum spanning tree by this procedure if no heuristic information influences the random walk. Using only heuristic information for constructing solutions, we show that a simple ACO algorithm together with the Broder-based construction procedure does not find a minimum spanning tree or even does not present a feasible solution in polynomial time.

After that, we consider a more incremental construction procedure that follows a general approach proposed by Dorigo and Stützle (2004) to obtain an ACO construction graph. We call this the Kruskal-based construction procedure as in each step an edge that does not create a cycle is chosen to be included into the solution. Using such a construction procedure, we are able to show the resulting algorithms are more efficient than simple evolutionary algorithms. Our analyses show how ACO algorithms for combinatorial optimization can be analyzed rigorously and are a first step in understanding ACO algorithms on more complicated structures. In particular, we provide insight into the working principles of ACO algorithms by studying the effect of the (guided) random walks of these algorithms.

Having motivated the analysis of stochastic search algorithms on the minimum spanning tree problem, we now give a survey on the rest of this chapter. In Section 5.1, we describe our model of the minimum spanning tree problem. The theory on minimum spanning trees is well established. In Section 5.2, we deduce some properties of local changes in non-optimal spanning trees which will be used in our analyses. In Section 5.3, we analyze evolutionary algorithms with respect to their computational complexity, and we study the impact of the construction graph for ACO algorithms in Section 5.4.

5.1 Representation for Evolutionary Algorithms

There are many ways to choose the search space for evolutionary algorithms when applying them to spanning tree problems. This problem has been investigated intensively by Raidl and Julstrom (2003). Their experiments point out that one should work with "edge sets". The search space equals $S = \{0, 1\}^m$, where each position corresponds to one edge. A search point $s \in S$ corresponds to the choice of all edges e_i, $1 \leq i \leq m$, where $s_i = 1$. The weight of edge e_i is denoted by w_i; $w_{\max} = \max_{1 \leq i \leq m} w_i$ and $w_{\min} = \min_{1 \leq i \leq m} w_i$ refer to the maximum and minimum weight of the given input graph. In many cases, many search points correspond to non-connected graphs and others correspond to connected graphs with cycles, i.e., graphs which are not trees. If all graphs which are not spanning trees have the same "bad" fitness, it will take exponential time to find a spanning tree when we apply a stochastic search algorithm. We will investigate two fitness functions f and f'.

Let

$$f(s) = (c(s), e(s) - (n - 1), w(s))$$

be the first fitness function, where $c(s)$ is the number of connected components of the graph described by s, and $e(s) = \sum_{i=1}^{n} s_i$ is the number of edges in this graph and $w(s) = \sum_{i=1}^{m} w_i s_i$ is the weight of the chosen edges. The fitness function has to be minimized with respect to the lexicographic order and takes the weight of all edges into account for which the corresponding bit $s_i = 1$ holds. The most important issue is to decrease $c(s)$ until we have graphs connecting all vertices. Then we have at least $n - 1$ edges, and the next issue is to decrease $e(s)$ under the condition that s describes a connected graph. Hence, we look for spanning trees. Finally, we look for minimum spanning trees.

It is necessary to penalize non-connected graphs since the empty graph has the smallest weight. However, it is not necessary to penalize extra connections

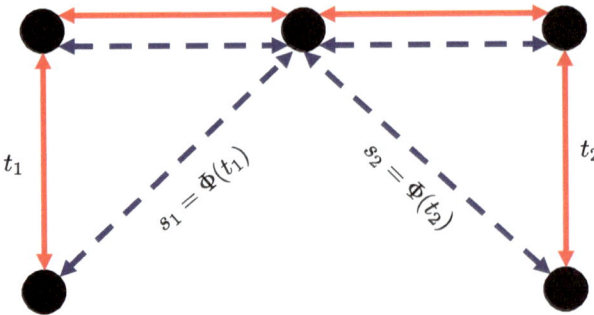

Fig. 5.1. Bijection. Continuous edges belong to a minimum spanning tree T, dashed edges correspond to a spanning tree S

since breaking a cycle decreases the weight. Therefore, it is also interesting to investigate the fitness function

$$f'(s) = (c(s), w(s)),$$

which should also be minimized with respect to the lexicographic order.

The fitness function f' is appropriate in the black-box scenario, which uses as little problem-specific knowledge as possible. The fitness function f contains the knowledge that optimal solutions are *trees*. This simplifies the analysis of stochastic search algorithms. Therefore, we always start with results on the fitness function f and discuss afterwards how to obtain similar results for f'.

5.2 Properties of Local Changes

The theory on minimum spanning trees is well established. Here we want to show how an arbitrary spanning tree can be turned into an optimal solution in a specific way that can be used later for analyzing the runtime of stochastic search algorithms. We identify a tree T by its set of edges. Let $e \in E \setminus T$ be an edge that is not contained in T. We denote by $Cyc(T, e)$ the edges of T that are contained in the cycle created when introducing e into T. We can construct from a spanning tree T another spanning tree T' by introducing an edge $e \in E \setminus T$ into T and removing one edge of $Cyc(T, e)$ from T. Such operations are called exchange operations. In this section, we recall some facts from the theory of minimum spanning trees that show that an arbitrary spanning tree T can be turned into an optimal solution T^* by a set of exchange operations where each operation is directly applicable on T and its execution of the operation does not lead to a weight increase. Using this, we can estimate the weight decrease possible when considering the current spanning tree T.

The following result was proven by Kano (1987) using an existence proof. Later, Mayr and Plaxton (1992) gave an explicit construction procedure, which we present in the following.

Theorem 5.1. *Let T be a minimum spanning tree and S be an arbitrary spanning tree of a given weighted graph $G = (V, E)$. Then there exists a bijection Φ from $T \setminus S$ to $S \setminus T$ such that for every edge $e \in T \setminus S$, $\Phi(e) \in Cyc(S, e)$ and $w(\Phi(e)) \geq w(e)$.*

Proof. Let C and D be disjoint subsets of E. The graph $G' = G[C, D]$ is constructed from G by contracting the edges of C and deleting the edges of D. We determine the bijection between the disjoint spanning trees $T' = T \setminus S$ and $S' = S \setminus T$ of the graph $G' = G[T \cap S, E \setminus T \setminus S]$. It is easy to see that $Cyc(T', e) \subseteq Cyc(T, e)$ holds for all $e \in T'$. Let t be the heaviest edge in T' and s be any edge in S' for which $t \in Cyc(T', s)$ and $s \in Cyc(S', t)$ holds. We can determine such an s by removing t from G'. This partitions the vertices of T' into two classes. Let s be the edge in S' that connects these two components. Note that $s \in Cyc(S', t)$ and $t \in Cyc(T', s)$ holds as s and t connect the two components of $T' \setminus \{t\}$.

T' is a minimum spanning tree of G', which implies that $w(t) \leq w(s)$. Set $\Phi(t) = s$ and determine the next component of the bijection by repeating the procedure on the graph $G'[s, t]$. $T'[s, t]$ is a minimum spanning tree of $G'[s, t]$. In addition, for all $e \in T'[s, t]$,

$$Cyc(S'[s, t], e) = Cyc(S', e) \setminus \{s\} \subseteq Cyc(S', e) \subseteq Cyc(S, e)$$

holds. Hence, the next assignment of an edge $e \in T'[s, t]$ to Φ will be guaranteed to satisfy $\Phi(e) \in Cyc(S, e)$. The process is iterated until for each $e \in T'$ a corresponding $\Phi(e) \in S'$ has been determined. $\square$

An illustration of the bijection is given in Figure 5.1. Note that the bijection gives a set of edge exchanges to transform an arbitrary spanning tree into a minimum spanning tree.

We denote by w_{opt} the weight of minimum spanning trees and want to show the following. For a search point s representing a non-minimum spanning tree, there are either many weight-decreasing local changes which, on average, decrease $f(s)$ by an amount that is not too small with respect to $w(s) - w_{\mathrm{opt}}$, or there are few of these local changes which, on the average, cause a larger decrease of the weight. This enables use to use the method of the expected multiplicative distance decrease presented in Section 4.2.3. Distance is in this case measured by the weight difference $w(s) - w_{\mathrm{opt}}$. The statement of the decrease by these local changes is made precise in the following lemma.

Lemma 5.2. *Let s be a search point describing a non-minimum spanning tree T. Then there exist some $k \in \{1, \ldots, n-1\}$ and k different accepted 2-bit flips such that the average distance decrease of these flips is at least $(w(s) - w_{\mathrm{opt}})/k$.*

Proof. Let s^* be a search point describing a minimum spanning tree T^*. Let $k := |T^* \setminus T|$. Then there exists a bijection $\Phi : T^* \setminus T \to T \setminus T^*$ such that

$\Phi(e)$ lies on the cycle $Cyc(T, e)$ and the weight of $\Phi(e)$ is not smaller than the weight of e due to Theorem 5.1.

We consider the k 2-bit flips flipping e and $\Phi(e)$ for $e \in T^* \setminus T$. They are accepted since e creates a cycle which is destroyed by the elimination of $\Phi(e)$. Performing all the k 2-bit flips simultaneously changes T to T^* and leads to a distance decrease of $w(s) - w_{\text{opt}}$. Hence, the average distance decrease of these steps is $(w(s) - w_{\text{opt}})/k$. $\square$

The analysis of stochastic search algorithms will be simplified if we can ensure that we always have the same parameter k in Lemma 5.2. This is easy if we allow also non-accepted 2-bit flips whose distance decrease is defined as 0. We add $n - k$ non-accepted 2-bit flips to the set of the k accepted 2-bit flips whose existence is proven in Lemma 5.2. Then we obtain a set of exactly n 2-bit flips. The total distance decrease is at least $w(s) - w_{\text{opt}}$ since this holds for the k accepted 2-bit flips. Therefore, the average distance decrease is bounded below by $(w(s) - w_{\text{opt}})/n$. We state this result as Lemma 5.3.

Lemma 5.3. *Let s be a search point describing a spanning tree T. Then there exists a set of n 2-bit flips such that the average distance decrease of these flips is at least $(w(s) - w_{\text{opt}})/n$.*

When analyzing the fitness function f' instead of f, we may accept non-spanning trees as improvements of spanning trees. Non-spanning trees can be improved by 1-bit flips eliminating edges of cycles. A 1-bit flip leading to a non-connected graph is not accepted and its distance decrease is defined as 0.

Lemma 5.4. *Let s be a search point describing a connected graph and consider the fitness function f'. Then there exist a set of n 2-bit flips and a set of $m - (n-1)$ 1-bit flips such that the average distance decrease of these flips is at least $(w(s) - w_{\text{opt}})/(m+1)$.*

Proof. We consider all 1-bit flips concerning the edges that are not contained in the minimum spanning tree T^*. If we try them in some arbitrary order we obtain a spanning tree T. If we consider their weight decrease with respect to the graph G' described by s, this weight decrease can be only larger. The reason is that a 1-bit flip, which is accepted in the considered sequence of 1-bit flips, is also accepted when applied to s. Then we apply Lemma 5.3 to T. At least the same weight decrease is possible by adding e_i and deleting a non-T^* edge with respect to G'. Altogether, we obtain at least a weight decrease of $w(s) - w_{\text{opt}}$. This proves the lemma, since we have chosen $m + 1$ flips. $\square$

5.3 Analysis of Evolutionary Algorithms

In this section, we analyze the computational complexity of evolutionary algorithms for computing a minimum spanning tree. We start by presenting upper bounds on the expected optimization time of $\mathrm{RLS}_b^{1,2}$ and $(1+1)$ EA_b. Afterwards, we present matching lower bounds and discuss how stochastic search algorithms can be sped up by using other mutation operators or parallelization.

5.3.1 Upper Bounds

The fitness function f penalizes solutions that are not connected or have more than $n-1$ edges. In the case of f', unconnected graphs are penalized. We first show that $\mathrm{RLS}_b^{1,2}$ and $(1+1)$ EA_b using the fitness functions f and f' construct connected graphs efficiently. In the case of f it is also easy to show that spanning trees are obtained in a small amount of time. We use the method of fitness-based partitions (see Section 4.2.1) and partition the search space into fitness levels with respect to the number of connected components or the number of edges for connected graphs.

Lemma 5.5. *The expected time until $\mathrm{RLS}_b^{1,2}$ or $(1+1)$ EA_b working on one of the fitness function f or f' has constructed a connected graph is $O(m \log n)$.*

Proof. The fitness functions are defined in such a way that the number of connected components will never be increased in accepted steps. For each edge set leading to a graph with k connected components, there are at least $k-1$ edges whose inclusion decreases the number of connected components by 1. Otherwise, the graph would not be connected. The probability of a step decreasing the number of connected components is at least $\frac{1}{2} \cdot \frac{k-1}{m}$ for $\mathrm{RLS}_b^{1,2}$ and $\frac{1}{e} \cdot \frac{k-1}{m}$ for $(1+1)$ EA_b. Hence, the expected time until s describes a connected graph is bounded above by

$$em \cdot \sum_{k=2}^{n} \frac{1}{k-1} = O(m \log n). \qquad \square$$

Lemma 5.6. *If s describes a connected graph, the expected time until $\mathrm{RLS}_b^{1,2}$ or $(1+1)$ EA_b constructs a spanning tree for the fitness function f is bounded by $O(m \log n)$.*

Proof. The fitness function f is defined in such a way that, starting with s, only connected graphs are accepted and that the number of edges does not increase. If s describes a graph with r edges, it contains a spanning tree with $n-1$ edges, and there are at least $r-(n-1)$ edges whose exclusion decreases the number of edges. If $r = n-1$, s describes a spanning tree. Otherwise, by the same arguments as in the proof of Lemma 5.5, we obtain an upper bound of

$$em \cdot \sum_{r=n}^{m} \frac{1}{r-(n-1)} = O(m \log(m-(n-1))) = O(m \log n). \qquad \square$$

This lemma holds also for $RLS_b^{1,2}$ and the fitness function f'. $RLS_b^{1,2}$ does not accept steps including only one edge or only two edges if s describes a connected graph. Since $RLS_b^{1,2}$ does not affect more than two edges in a step, it does not accept steps in which the number of edges of a connected graph is increased. This does not hold for $(1+1)$ EA_b. It is possible that the exclusion of one edge and the inclusion of two or more edges creates a connected graph whose weight is not larger than the weight of the given graph.

In the following, we prove an upper bound of size $O(m^2(\log n + \log w_{\max}))$ on the expected optimization time for arbitrary graphs using the method of expected multiplicative distance decrease (see Section 4.2.3). This bound relies on the properties of minimum spanning trees, which we have stated in Section 5.1 and is $O(m^2 \log n)$ as long as $w_{\max}$ is polynomially bounded. But it is always polynomially bounded with respect to the bit length of the input. Theorem 5.9 shows that the bound is optimal.

Theorem 5.7. *The expected time until $RLS_b^{1,2}$ or $(1+1)$ EA_b working on the fitness function f constructs a minimum spanning tree is bounded by $O(m^2(\log n + \log w_{\max}))$.*

Proof. By Lemmas 5.5 and 5.6, it is sufficient to investigate the search process after finding a search point s describing a spanning T. Then, by Lemma 5.3, there always exists a set of n 2-bit flips whose average distance decrease is at least $(w(s) - w_{\mathrm{opt}})/n$. The choice of such a 2-bit flip is called a "good step". The probability of performing a good step equals $\Theta(n/m^2)$ and each good step is chosen with the same probability. A good step decreases the difference between the weight of the current spanning tree and w_{opt} on average by a factor not larger than $1 - 1/n$. This holds independently of previous good steps. Hence, after N good steps, the expected difference in the weight of T and w_{opt} is bounded above by $(1 - 1/n)^N \cdot (w(s) - w_{\mathrm{opt}})$. Since $w(s) \le (n-1) \cdot w_{\max}$ and $w_{\mathrm{opt}} \ge 0$, we obtain the upper bound $(1 - 1/n)^N \cdot D$, where $D := (n-1) \cdot w_{\max}$.

If $N := \lceil (\ln 2) \cdot (n-1) \cdot (\log D + 1) \rceil$, this bound is at most $\frac{1}{2}$. Since the difference is not negative, by Markov's inequality, the probability that the bound is less than 1 is at least $1/2$. The difference is an integer implying that the probability of finding a minimum spanning tree is at least $1/2$. Repeating the same arguments, the expected number of good steps until a minimum spanning tree is found is bounded by $2N = O(n \log D) = O(n(\log n + \log w_{\max}))$.

By our construction, there are always exactly n good 2-bit flips. Therefore, the probability of a good step does not depend on the current search point. Hence, the expected time until r steps are good equals $\Theta(rm^2/n)$. Altogether, the expected optimization time is bounded by

$$O(Nm^2/n) = O(m^2(\log n + \log w_{\max})). \qquad \square$$

Applying Lemma 5.4 instead of Lemma 5.3, it is not too difficult to obtain the same upper bound for the fitness function f'. The main difference is that a good 1-bit flip has a larger probability than a good 2-bit flip.

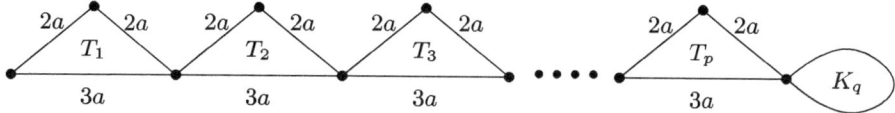

Fig. 5.2. Example graph TG with p connected triangles and a complete graph on q vertices with edges of weight 1

Theorem 5.8. *The expected time until* $RLS_b^{1,2}$ *or (1+1)* EA_b *working on the fitness function* f' *constructs a minimum spanning tree is bounded by* $O(m^2(\log n + \log w_{\max}))$.

Proof. By Lemma 5.5, it is sufficient to analyze the phase after having constructed a connected graph. We apply Lemma 5.4. The total distance decrease of the chosen 1-bit flips and 2-bit flips is at least $w(s) - w_{\mathrm{opt}}$ if s is the current search point. If the total distance decrease of the 1-bit flips is larger than the total distance decrease of the chosen 2-bit flips, the step is called a 1-step. Otherwise, it is called a 2-step.

   If more than half of the steps are 2-steps, we adapt the proof of Theorem 5.7 with $N' := 2N$ since we guarantee only an expected distance decrease by a factor of $1 - 1/(2n)$. Otherwise, we consider the good 1-steps which have an expected weight decrease of a factor of $1 - 1/(2m')$ for $m' = m - (n - 1)$. Choosing $M := \lceil 2 \cdot (\ln 2) \cdot m' \cdot (\log D + 1) \rceil$, we can apply the proof technique of Theorem 5.7, where M plays the role of N. The probability of performing a good 1-bit flip equals $\Theta(m'/m)$. In this case, we obtain the bound

$$O(Mm/m') = O(m(\log n + \log w_{\max}))$$

for the expected number of steps, which is even smaller than the proposed bound. □

5.3.2 Lower Bound

After having given upper bounds we show lower bounds on the expected optimization time. To do this, we investigate the example graph TG shown in Figure 5.2. The graph TG consists of a connected sequence of p triangles and the last triangle is connected to a complete graph on q vertices. The number of vertices equals $n := 2p+q$ and the number of edges equals $m := 3p+q(q-1)/2$. We consider the case of $p = n/4$ and $q = n/2$ implying that $m = \Theta(n^2)$. The edges in the complete graph have the weight 1 and we set $a := n^2$. Each triangle edge has a weight which is larger than the weight of all edges of the complete graph altogether. Theorems 5.7 and 5.9 prove that this graph is a worst-case instance with polynomial weights.

Theorem 5.9. *The expected optimization time until* $RLS_b^{1,2}$ *and (1+1)* EA_b *find a minimum spanning tree for the example graph* TG *equals* $\Theta(m^2 \log n) = \Theta(n^4 \log n)$ *with respect to the fitness functions* f *and* f'.

Proof. The upper bounds are contained in Theorems 5.9 and 5.7. Here we prove the lower bound by investigating typical runs of the algorithm. We partition the graph TG into its triangle part T and its clique part C. Each search point x describes an edge set. We use the following notation:

- $d(x)$: number of disconnected triangles with respect to the edges chosen by x,
- $b(x)$: number of bad triangles (exactly one $2a$-edge and the $3a$-edge are chosen),
- $g(x)$: number of good triangles (exactly the two $2a$-edges are chosen),
- $c(x)$: number of complete triangles (all three edges are chosen),
- $\mathrm{con}_G(x)$: number of connected components in the graph,
- $\mathrm{con}_C(x)$: number of connected components in the clique part C of the graph,
- $\mathrm{con}_T(x)$: number of connected components in the tree part T of the graph.

We investigate four phases of the search. The first phase of length 1 is the initialization step producing the random edge set x. In the following, all statements hold with probability $1 - o(1)$.

Claim. After initialization, $b(x) = \Theta(n)$ and $\mathrm{con}_C(x) = 1$.

Proof. The statements can be proved independently since the corresponding parts of x are created independently. The probability that a given triangle is bad equals $1/4$. There are $n/4$ triangles and $b(x) = \Theta(n)$ by Chernoff bounds. We consider one vertex of C. It has $n/2 - 1$ possible neighbors. By Chernoff bounds, it is connected to at least $n/6$ of these vertices. For each other vertex, the probability of not being connected to at least one of these $n/6$ vertices is $(1/2)^{n/6}$. This is unlikely even for one of the remaining vertices. Hence, $\mathrm{con}_C(x) = 1$. □

For the following phases, we distinguish between the steps by the number k of flipping triangle edges and call them k-steps. Let p_k be the probability of a k-step. For $\mathrm{RLS}_b^{1,2}$, $p_1 = \Theta(n^{-1})$, $p_2 = \Theta(n^{-2})$ and $p_k = 0$, if $k \geq 3$. For $(1+1)$ EA_b and constant k

$$p_k = \binom{3n/4}{k} \left(\frac{1}{m}\right)^k \left(1 - \frac{1}{m}\right)^{3n/4-k} = \Theta(n^k m^{-k}) = \Theta(n^{-k}).$$

For a phase of length $n^{5/2}$, the following statements hold. The number of 1-steps equals $\Theta(n^{3/2})$, the number of 2-steps equals $\Theta(n^{1/2})$, and there is no k-step, $k \geq 3$.

Claim. Let $b(x) = \Theta(n)$ and $\mathrm{con}_C(x) = 1$. In a phase of length $n^{5/2}$, a search point y where $b(y) = \Theta(n)$ and $\mathrm{con}_G(y) = 1$ is produced.

Proof. By Lemma 5.5, the probability of creating a connected graph is large enough. Let y be the first search point where $\mathrm{con}_G(y) = 1$. We prove that $b(y) = \Theta(n)$. All the 2-steps can decrease the b-value by at most $O(n^{1/2})$. A 1-step has two ways to destroy a bad triangle.

- It may destroy an edge of a bad triangle. This increases the con_G-value. In order to accept the step, it is necessary to decrease the con_C-value.
- It may add the missing edge to a bad triangle. This increases the weight by at least $2a$. No triangle edge is eliminated in this step. In order to accept the step, it is necessary to decrease the con_C-value.

However, $\mathrm{con}_C(x) = 1$. In order to decrease this value, it has to be increased before. A step increasing the con_C-value can be accepted only if the con_T-value is decreased in the same step at least by the same amount. This implies that triangle edges have to be added. For a 1-step, the total weight is increased without decreasing the con_G-value and the step is not accepted. Hence, only the $O(n^{1/2})$ 2-steps can increase the con_C-value. By Chernoff bounds, the number of clique edges flipping in these steps is $O(n^{1/2})$. This implies that the number of bad triangles is decreased by only $O(n^{1/2})$. □

Claim. Let $b(y) = \Theta(n)$ and $\mathrm{con}_G(y) = 1$. In a phase of length $n^{5/2}$, a search point z where $b(z) = \Theta(n)$, $\mathrm{con}_G(z) = 1$, and $T(z)$ is a tree is produced.

Proof. Only search points x describing connected graphs are accepted, in particular, $d(x) = 0$. Let z be the first search point where $T(z)$ is a tree. Then $\mathrm{con}_G(z) = 1$ and we have to prove that $b(z) = \Theta(n)$ and that z is produced within $n^{5/2}$ steps. A 1-step can be accepted only if it turns a complete triangle into a good or bad triangle. Such a step is accepted if no other edge flips. Moreover, $c(x)$ cannot be increased. In order to increase $c(x)$ it is necessary to add the missing edge to a good or bad triangle. To compensate for this weight increase, we have to eliminate an edge of a complete triangle. Remember that we have no k-steps for $k \geq 3$. If $c(x) = l$, the probability of decreasing the c-value is at least $3l/(em)$ and the expected time to eliminate all complete triangles is $O(m \log n) = O(n^2 \log n)$. Hence, $n^{5/2}$ steps are sufficient to create z. The number of bad triangles can be decreased only in the $O(n^{1/2})$ 2-steps implying that $b(z) = \Theta(n)$. □

Claim. Let $b(z) = \Theta(n)$, $\mathrm{con}_G(z) = 1$, and $T(z)$ be a tree. The expected time of finding a minimum spanning tree is $\Omega(n^4 \log n)$.

Proof. First, we assume that only 2-steps change the number of bad triangles. Later, we complete the arguments. The expected waiting time for a 2-step flipping those two edges of a bad triangle that turn it into a good one equals $\Theta(n^4)$. The expected time to decrease the number of bad triangles from b to $b - 1$ equals $\Theta(n^4/b)$. Since b has to be decreased from $\Theta(n)$ to 0, we obtain an expected waiting time of

$$\Theta\left(n^4 \cdot \sum_{1 \le b \le \Theta(n)} (1/b)\right) = \Theta(n^4 \log n). \tag{$*$}$$

Similarly to the coupon collector's theorem (see Appendix A.13) we obtain that the optimization step if only 2-steps can be accepted equals $\Theta(n^4 \log n)$ with probability $1 - o(1)$. Hence, it is sufficient to limit the influence of all k-steps, $k \ne 2$, to a time period of $\alpha n^4 \log n$ for some constant $\alpha > 0$. Again with probability $1 - o(1)$, the number of 4-steps is $O(\log n)$ and there are no k-steps for $k \ge 5$. The 4-steps can decrease the number of bad triangles by at most $O(\log n)$. Because of the weight increase, a k-step, $k \le 4$, can be accepted only if it eliminates at least $\lceil k/2 \rceil$ triangle edges. Moreover, it is not possible to disconnect a good or a bad triangle. Hence, a 4-step cannot create a complete triangle. As long as there is no complete triangle, a 3-step or a 1-step has to disconnect a triangle and is not accepted. A 2-step can only be accepted if it changes a bad triangle into a good one. Hence, no complete triangles are created. The 4-steps eliminate $O(\log n)$ terms of the sum in $(*)$. The largest terms are those for the smallest values of b. We only have to subtract a term of $O(n^4 \log \log n) = o(n^4 \log n)$ from the bound $\Theta(n^4 \log n)$, and this proves the claim. $\square$

This completes the proof since the sum of all failure probabilities is $o(1)$. $\square$

5.3.3 Speed-Up Techniques

Theorems 5.9, 5.7, and 5.8 contain matching upper and lower bounds for $\text{RLS}_b^{1,2}$ and $(1+1)$ EA_b with respect to the fitness functions f and f'. The bounds are worst-case bounds and one can hope that the algorithms are more efficient for many graphs. Here we discuss what can be gained by other evolutionary algorithms.

First, we introduce more problem-specific mutation operators. It is easy to construct spanning trees. Afterwards, it is good to create children with the same number of edges. The new mutation operators are:

- If $\text{RLS}_b^{1,2}$ flips two bits, it chooses randomly a 0-bit and a 1-bit.
- If s contains k 1-bits, $(1+1)$ EA_b flips each 1-bit with probability $1/k$ and each 0-bit with probability $1/(m - k)$.

Jansen and Sudholt (2005) have analyzed this mutation operator in greater detail. One important result of their work is that simple but not trivial pseudo-boolean functions can be optimized by evolutionary algorithms in time $O(n)$, which breaks for the first time the $\Theta(n \log n)$ bound that is often the result of the analysis on simple pseudo-Boolean functions.

For spanning trees, the probability of a specific edge exchange is increased from $\Theta(1/m^2)$ to $\Theta(1/(n(m - n + 1)))$. The following result can be obtained by adjusting the proofs to the modified mutation operators.

Theorem 5.10. *For the modified mutation operators, the bounds of Theorems 5.7, 5.8, and 5.9 can be replaced by bounds of size $\Theta(mn \log n)$ and $O(mn(\log n + \log w_{\max}))$ respectively.*

When using larger populations, we have to pay for improving all members of the population. This holds at least if we guarantee a large diversity in the population. The lower bound of Theorem 5.9 holds with overwhelming probability. Hence, we do not expect that large populations help. The analysis in the proof of Theorems 5.7 and 5.8 is quite precise in most aspects. There is only one essential exception. We know that the weight distance to w_{opt} is decreased on average by a factor of at most $1 - 1/n$ and we work under the pessimistic assumption that this factor equals $1 - 1/n$. For large populations or multi-starts the probability of having sometimes much larger improvements may increase for many graphs.

It is more interesting to "parallelize" the algorithms by producing more children in parallel. The algorithm $(1+\lambda)$ EA$_b$ differs from $(1+1)$ EA$_b$ by producing in each iteration independently λ children from the single individual of the current population. The selection procedure selects an individual with the smallest f-value (or f'-value) from among the parent and its children. In a similar way, we obtain λ-PRLS$_b^{1,2}$ (parallel RLS$_b^{1,2}$) from RLS$_b^{1,2}$. In the proofs of Theorems 5.7 and 5.8, we have seen that the probability of a good step is $\Theta(n/m^2)$. Choosing $\lambda = \lceil m^2/n \rceil$, this probability is increased to a positive constant. We have seen that the expected number of good steps is bounded by $O(n(\log n + \log w_{\max}))$. This leads to the following result.

Theorem 5.11. *The expected number of generations until λ-PRLS$_b^{1,2}$ or $(1+\lambda)$ EA$_b$ with $\lambda := \lceil m^2/n \rceil$ children constructs a minimum spanning tree is $O(n(\log n + \log w_{\max}))$. This holds for the fitness functions f and f'.*

If we use the modified mutation operator defined above, the probability of a good step is $O(1/m)$ and we obtain the same bound on the expected number of generations as in Theorem 5.11 for $\lambda := m$.

Crossover operators are considered important in evolutionary computation. But one-point crossover or two-point crossover is not appropriate for edge set representations. It is not possible to build blocks of all edges adjacent to a vertex. For uniform crossover, it is very likely to create graphs which are not spanning trees. Hence, only problem-specific crossover operators seem to be useful. Such operators are described by Raidl and Julstrom (2003). It is difficult to analyze stochastic search algorithms with these crossover operators. Up to now, no results which prove better runtime bounds than the ones presented in this chapter have been obtained for such algorithms.

5.4 Analysis of Ant Colony Optimization

In this section, we investigate the computational complexity of ACO algorithms for the computation of minimum spanning trees. We focus in partic-

---

Algorithm 7 MMAS

Set $\tau_{(u,v)} = 1/|A|$ for all $(u, v) \in A$.
Compute a solution x using a construction procedure.
Update the pheromone values and set $x^* := x$.
repeat
    Compute x using a construction procedure.
    if $f(x) < f(x^*)$ **then**
        set $x^* := x$.
    end if
    update the pheromone values.
until stop

---

ular on the impact of the chosen construction graph. Formulating the MST problem as a problem of pseudo-Boolean optimization with the fitness functions presented in the previous sections, similar results as in the previous section can be obtained. This is due to the fact that simple ACO algorithms for pseudo-Boolean optimization behave like $(1+1)$ EA_b when choosing a certain parameter setting (Neumann and Witt, 2009). Hence, many results on $(1+1)$ EA_b for combinatorial optimization problems can be transfered to ACO algorithms in this scenario. However, it is more natural to investigate construction graphs which are more related to the given problem. As the MST problem is a graph problem, it seems natural to take the input graph as a construction graph into account.

We study a variant of the Max-Min Ant System (MMAS) introduced by Stützle and Hoos (2000). In our MMAS, solutions are constructed iteratively by different construction procedures on a given directed construction graph $C = (X, A)$. In the initialization step, each edge $(u, v) \in A$ gets a pheromone value $\tau_{(u,v)} = 1/|A|$ such that the pheromone values sum up to 1. Afterwards, an initial solution x^* is produced by a random walk of an imaginary ant on the construction graph and the pheromone values are updated with respect to this walk. In each iteration, a new solution is constructed and the pheromone values are updated if this solution is not inferior (with respect to a fitness function f) to the best solution obtained so far.

Our construction procedures construct in each iteration a tree T of the given graph. Therefore, the fitness of a solution is given by the weight of the edges contained in T. We consider the expected number of solutions that are constructed by the algorithm until a minimum spanning tree has been obtained for the first time. We call this the *expected optimization time* of the MMAS.

5.4.1 Broder-Based Construction Graph

Since the MST problem is a graph problem, the first idea is to use the input graph G to the MST problem itself as the construction graph C of the MMAS.

Algorithm 8 BroderConstruct(G, τ, η)

Choose an arbitrary node $s \in V$.

$u := s$, $T = \emptyset$.

while not all nodes of G have been visited **do**

Let $R := \sum_{\{u,v\} \in E} [\tau_{\{u,v\}}]^\alpha \cdot [\eta_{\{u,v\}}]^\beta$.

Choose neighbor v of u with probability $\frac{[\tau_{\{u,v\}}]^\alpha \cdot [\eta_{\{u,v\}}]^\beta}{R}$.

if v has not been visited before **then**

set $T := T \cup \{u,v\}$.

end if

Set $u := v$.

end while

Return T.

(Note that each undirected edge $\{u,v\}$ can be considered as two directed edges (u,v) and (v,u).) However, it is not obvious how a random walk of an ant on G is translated into a spanning tree. Interestingly, the famous algorithm of Broder (1989), which chooses uniformly at random from all spanning trees of G, is a random walk algorithm.

We will use an ACO variant of Broder's algorithm as given in Algorithm 8. As usual in ACO algorithms, the construction procedure maintains pheromone values τ and heuristic information η for all edges of the construction graph G. In the MST problem, we assume that the heuristic information $\eta_{\{u,v\}}$ of an edge $\{u,v\}$ is the inverse of the weight of the edge $\{u,v\}$ in G. α and β are parameters that control the extent to which pheromone values and heuristic information is used.

Obviously, Algorithm 8 outputs a spanning tree T, whose cost $f(T)$ is measured by the sum of the w-values of its edges. After a new solution has been accepted, the pheromone values τ are updated with respect to the constructed spanning tree T. We maintain upper and lower bounds on these values, which are common measures to ensure convergence (Dorigo and Blum, 2005). We assume that after each update, the τ-value of each edge in the construction graph attains either the upper bound h or the lower bound ℓ. Hence, for the new pheromone values τ' after an update, it holds that

$$\tau'_{\{u,v\}} = h \quad \text{if} \quad \{u,v\} \in T \quad \text{and} \quad \tau'_{\{u,v\}} = \ell \quad \text{if} \quad \{u,v\} \notin T.$$

So the last constructed solution is indirectly saved by the $n-1$ undirected edges that obtain the high pheromone value h. The ratio of the parameters ℓ and h is crucial since too large values of ℓ will lead to too large changes of the tree in subsequent steps whereas too large values of h will make changes of the tree too unlikely. We choose h and ℓ such that $h = n^3 \ell$ holds and will argue later on the optimality of this choice.

Note that choosing $\beta = 0$ or $\alpha = 0$ in Algorithm 8, only the pheromone value or the heuristic information influences the random walk. We examine the cases where one of these values is 0 to study the effect of the pheromone values or the heuristic information separately. First, we consider the case $\alpha = 1$ and $\beta = 0$ for the Broder-based construction graph. This has the following consequences. Let u be the current node of the random walk and denote by $R := \sum_{\{u,v\} \in E} \tau_{\{u,v\}}$ the sum over the pheromone values of all edges that are incident on u. Then the next node is chosen proportionally to the pheromone values on the corresponding edges, which means that a neighbor v of u is chosen with probability $\tau_{\{u,v\}}/R$.

For simplicity, we call the described setting of α, β, h, and ℓ the *cubic update scheme*. To become acquainted therewith, we derive the following simple estimations on the probabilities of traversing edges depending on the pheromone values. Assume that a node v has k adjacent edges with value h and i adjacent edges with value ℓ. Note that $k + i \leq n - 1$ and $h = n^3\ell$. Then the probability of choosing an edge with value h is

$$\frac{kh}{kh + i\ell} = 1 - \frac{i}{kn^3 + i} \geq 1 - \frac{1}{n^2},$$

where among the edges with values h one edge is chosen uniformly at random. The probability of choosing a specific edge with value ℓ is at least

$$\frac{\ell}{\ell + (n-2)h} \geq \frac{\ell}{nh} = \frac{1}{n^4}.$$

This leads us to the following theorem, which shows that the MMAS in the described setting is able to construct MSTs in expected polynomial time.

Theorem 5.12. *The expected optimization time of the MMAS using the procedure* BroderConstruct *with cubic update scheme is $O(n^6(\log n + \log w_{\max}))$. The expected number of traversed edges in a run of* BroderConstruct *is bounded above by $O(n^2)$ except for the initial run, where it is $O(n^3)$.*

Proof. We use the following idea of Theorem 5.1. Suppose the spanning tree T^* was constructed in the last accepted solution. Let $T = T^* \setminus \{e\} \cup \{e'\}$ be any spanning tree that is obtained from T^* by including one edge e' and removing another edge e, and let $s(m,n)$ be a lower bound on the probability of producing T from T^* in the next step. Then the expected number of steps until a minimum spanning tree has been obtained is $O(s(m,n)^{-1}(\log n + \log w_{\max}))$. To prove the theorem, it therefore suffices to show that the probability of the MMAS producing T by the next constructed solution is $\Omega(1/n^6)$.

To simplify our argumentation, we first concentrate on the probability of rediscovering T^* in the next constructed solution. This happens if the ant traverses all edges of T^* in some arbitrary order and no other edges in between, which might require that an edge has to be taken more than once. (This is a pessimistic assumption since newly traversed edges are not necessarily included in the solution.) Hence, we are confronted with the cover time for the tree T^*. The cover time for trees on n nodes in general is bounded above by $2n^2$ (Motwani and Raghavan, 1995), i.e., by Markov's inquality, it is at most

$4n^2$ with probability at least $1/2$. We can apply this result if no so-called error occurs such that an edge with pheromone value ℓ is taken. According to the above calculations, the probability of an error is bounded above by $1/n^2$ in a single step of the ant. Hence, there is no error in $O(n^2)$ steps with probability $\Omega(1)$. Therefore, the probability of rediscovering T^* in the next solution (using $O(n^2)$ steps of *BroderConstruct*) is at least $\Omega(1)$. Additionally, taking into account the number of steps $O(n^3)$ for the initial solution (Broder, 1989), we have already bounded the expected number of traversed edges in a run of *BroderConstruct*.

To construct T instead of T^*, exactly one error is desired, namely e' has to be traversed instead of e. Consider the ant when it is for the first time on a node on which e' is incident. By the calculations above, the probability of including e' is $\Omega(1/n^4)$. Note that inserting e' into T^* closes a cycle c. Hence, when e' has been included, there may be at most $n-2$ edges of $\tilde{T} := T^* \setminus \{e\}$ left to traverse. We partition the edges of the forest $\tilde{T}$ into two subsets: The edges that belong to the cycle c are called critical and the remaining ones are called uncritical. The order of inclusion for the uncritical edges is irrelevant. However, all critical edges have to be included before the ant traverses edge e.

We are faced with the following problem: Let $v_1, \ldots, v_k, v_1$ describe the cycle c and suppose w. l. o. g. that $e' = \{v_1, v_k\}$. It holds that $e = \{v_i, v_{i+1}\}$ for some $1 \le i \le k-1$. Moreover, let v_s be the node of c that is visited first by the ant. W. l. o. g., $1 \le s \le i$. With probability $\Omega(1/n^4)$, the edge e' is traversed exactly once until a new solution has been constructed. Hence, after e' has been taken, the ant must visit the nodes $v_k, v_{k-1}, \ldots, v_{i+1}$ in the described order (unless an error other than including e' occurs), possibly traversing uncritical edges in between. To ensure that e has not been traversed before, we would like the ant to visit all the nodes in $\{v_1, \ldots, v_i\}$, without visiting nodes in $\{v_{i+1}, \ldots, v_k\}$, before visiting v_k by traversing e'. We apply results from the fair gambler's ruin problem given in Section 4.2.5. The probability of going from v_s to v_i before visiting v_k is at least $\Omega(1/n)$. The same lower bound holds on the probability of going from v_i to v_1 before visiting v_{i+1}. These random walks are still completed in expected time $O(n^2)$. Hence, in total, the probability of constructing T is $\Omega((1/n^4) \cdot (1/n) \cdot (1/n)) = \Omega(1/n^6)$, as suggested. $\square$

We see that the ratio $h/\ell = n^3$ leads to relatively large exponents in the expected optimization time. However, this ratio seems to be necessary for our argumentation. Consider the complete graph on n nodes where the spanning tree T^* equals a path of length $n-1$. The cover time for this special tree T^* is bounded below by $\Omega(n^2)$. To each node of the path, at most two edges with value h and at least $n-3$ edges with value ℓ are incident. Hence, the ratio is required to obtain an error probability of $O(1/n^2)$. It is much more difficult to improve the upper bound of Theorem 5.12 or to come up with a matching lower bound. The reasons are two. First, we cannot control the effects of steps where the ant traverses edges to nodes that have been visited before in the construction step. These steps might reduce the time until certain edges of T^* are reached. Second, our argumentation concerning the cycle $v_1, \ldots, v_k, v_1$ makes a worst-case assumption on the starting node v_s. It seems more likely that v_s is uniform over the path, which could improve the upper bound of the theorem by a factor $\Omega(n)$. However, a formal proof of this is open.

ACO algorithms often use heuristic information to direct the search process. In the following, we set $\alpha = 0$ and examine the effect of heuristic information for the MST problem. Recall that the heuristic information for an edge e is given by $\eta(e) = 1/w(e)$. Interestingly, for the obvious Broder-based graph, heuristic information alone does not help us find MSTs in reasonable time, regardless of β. On the following example graph G^* (see Figure 5.3), either the runtime of *BroderConstruct* explodes or MSTs are found only with exponentially small probability. W.l.o.g., $n = 4k + 1$. Then G^*, a connected graph on the nodes $\{1, \ldots, n\}$, consists of k triangles with weights $(1, 1, 2)$ and two paths of length k with exponentially increasing weights along the path. More precisely, let

$$T^* := \bigcup_{i=1}^{k} \{\{1, 2i\}, \{1, 2i+1\}, \{2i, 2i+1\}\},$$

where $w(\{1, 2i\}) = w(\{2i, 2i+1\}) := 1$ and $w(\{1, 2i+1\}) := 2$. Moreover, denote

$$P_1^* := \{1, 2k+2\} \cup \bigcup_{i=2}^{k} \{2k+i, 2k+i+1\},$$

where $w(\{1, 2k+2\}) := 2$ and $w(\{2k+i, 2k+i+1\}) := 2^i$, and, similarly,

$$P_2^* := \{1, 3k+2\} \cup \bigcup_{i=2}^{k} \{3k+i, 3k+i+1\},$$

where $w(\{1, 3k+2\}) := 2$ and $w(\{3k+i, 3k+i+1\}) := 2^i$. Finally, the edge set of G^* is $T^* \cup P_1^* \cup P_2^*$. Hence, all triangles and one end of each path are glued by node 1.

Theorem 5.13. *Choosing $\alpha = 0$ and β arbitrarily, the probability that the MMAS using* BroderConstruct *finds an MST for G^*, or the probability of termination within polynomial time, is $2^{-\Omega(n)}$.*

Proof. Regardless of the ant's starting point, at least one path, w.l.o.g. P_1^*, must be traversed from noded 1 to its other end, and for least $k-1$ triangles, both nodes $2i$ and $2i+1$ must be visited through node 1. For each of these initially undiscovered triangles, the first move into the triangle must go from 1 to $2i$, otherwise the resulting tree will not be minimal. If the triangle is entered at node $2i$, we consider it a success, and otherwise (entrance at $2i+1$) an error. The proof idea is to show that for too small β, i.e., when the influence

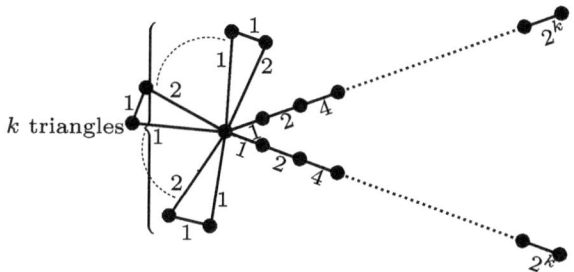

Fig. 5.3. Graph G^* consisting of k triangles and two paths of length k

of heuristic information is low, with overwhelming probability at least one triangle contains an error. If, on the other hand, β is too large, the ant with overwhelming probability will not be able to traverse P_1^* in polynomial time due to its exponentially increasing edge weights.

We study the success probabilities for the triangles and the path P_1. Given that the ant moves from 1 to either $2i$ or $2i + 1$, the probability of going to $2i$ equals

$$\frac{(\eta(\{1, 2i\}))^\beta}{(\eta(\{1, 2i\}))^\beta + (\eta(\{1, 2i + 1\}))^\beta} = \frac{1}{1 + 2^{-\beta}}$$

since $\eta(e) = 1/w(e)$. Therefore, the probability of $k - 1$ successes equals, due to independence, $(1 + 2^{-\beta})^{-k+1}$. This probability increases with β. However, for $\beta \leq 1$, it is still bounded above by $(2/3)^{k-1} = 2^{-\Omega(n)}$.

Considering the path P_1^*, we are faced with the unfair gambler's ruin problem (see Section 4.2.5). At each of the nodes $2k + i$, $2 \leq i \leq k - 1$, the probability of going to a lower-numbered node and the probability of going to a higher-numbered one have the same ratio of $r := (2^{-i+1})^\beta/(2^i)^\beta = 2^\beta$. Hence, starting in $2k + 2$, the probability of reaching $3k + 1$ before returning to 1 equals $\frac{r}{r^k - 1} = \frac{2^\beta}{2^{k\beta} - 1}$. This probability decreases with β. However, for $\beta \geq 1$, it is still bounded above by $2/(2^k - 1) = 2^{-\Omega(n)}$. Then the probability of reaching the end in a polynomial number of trials is also $2^{-\Omega(n)}$. □

5.4.2 A Kruskal-Based Construction Procedure

Dorigo and Stützle (2004) present a general approach on how to obtain an ACO construction graph from any combinatorial optimization algorithm. The idea is to identify the so-called components of the problem, which may be objects, binary variables etc., with nodes of the construction graph and to allow the ant to choose from these components by moving to the corresponding nodes. In our setting, the components to choose from are the edges from the edge set $\{1, \ldots, m\}$ of the input graph G. Hence, the canonical construction graph $C(G)$ for the MST problem is a directed graph on the $m + 1$ nodes

---

Algorithm 9 Construct$(C(G), \tau, \eta)$

$v_0 := s;\ k := 0$.
while $N(v_k) \neq \emptyset$ **do**
  Let $R := \sum_{y \in N(v_k)} [\tau_{(v_k, y)}]^\alpha \cdot [\eta_{(v_k, y)}]^\beta$.
  Choose neighbor $v_{k+1} \in N(v_k)$ with probability $\frac{[\tau_{(v_k, v_{k+1})}]^\alpha \cdot [\eta_{(v_k, v_{k+1})}]^\beta}{R}$.
  Set $k := k + 1$.
end while
Return the path $p = (v_0, \ldots, v_k)$.

$\{0, 1, \ldots, m\}$ with the designated start node $s := 0$. Its edge set A of cardinality m^2 is given by

$$A := \big\{(i, j) \mid 0 \leq i \leq m,\ 1 \leq j \leq m,\ i \neq j\big\},$$

i.e., $C(G)$ is obtained from the complete directed graph by removing all self-loops and the edges pointing to s. When the MMAS visits node e in the construction graph $C(G)$, this corresponds to choosing the edge e for a spanning tree. To ensure that a walk of the MMAS actually constructs a tree, we define the feasible neigborhood $N(v_k)$ of node v_k depending on the nodes $v_1, \ldots, v_k$ visited so far:

$$N(v_k) := \big(E \setminus \{v_1, \ldots, v_k\}\big) \setminus \ \big\{e \in E \mid (V, \{v_1, \ldots, v_k, e\}) \text{ contains a cycle}\big\}.$$

Note that the feasible neighborhood depends on the memory of the ant about the path followed so far, which is very common in ACO algorithms (Dorigo and Stützle, 2004).

A new solution is constructed using Algorithm 9. Again, the random walk of an ant is controlled by the pheromone values τ and the heuristic information η on the edges. As in the Broder-based construction graph, we assume that the $\eta_{(u,v)}$-value of an edge (u, v) is the inverse of the weight of the edge of G corresponding to the node v in $C(G)$.

A run of Algorithm 9 returns a sequence of $k + 1$ nodes of $C(G)$. It is easy to see that $k := n - 1$ after the run. Hence the number of steps is bounded above by n, and $v_1, \ldots, v_{n-1}$ is a sequence of edges that form a spanning tree for G. Accordingly, we measure the fitness $f(p)$ of a path $p = (v_0, \ldots, v_{n-1})$ simply by $w(v_1) + \cdots + w(v_{n-1})$, i.e., the cost of the corresponding spanning tree. It remains to specify the update scheme for the pheromone values. As in the case of the Broder-based construction procedure, we only consider two different values h and ℓ. To allow the ant to rediscover the edges of the previous spanning tree equiprobably in each order, we reward all edges pointing to nodes from p except s, i.e., we reward $(m + 1)(n - 1)$ edges. Hence, the τ'-values are

$$\tau'_{(u,v)} = h \quad \text{if } v \in p \text{ and } v \neq s \quad \text{and} \quad \tau'_{(u,v)} = \ell \quad \text{otherwise}.$$

We choose h and ℓ such that $h = (m - n + 1)(\log n)\ell$ holds. In this case, the probability of taking a rewarded edge (if applicable) is always at least $1 - 1/\log n$.

We consider the case where the random walk to construct solutions is only influenced by the pheromone values on the edges of $C(G)$. The following result can be obtained by showing that the probability of obtaining from the current tree T^* a tree $T = T^* \setminus \{e\} \cup \{e'\}$ is lower bounded by $\Omega(1/(mn))$. The proof can be carried out in a similar fashion as done for Theorem 5.12.

Theorem 5.14. *Choosing $\alpha = 1$ and $\beta = 0$, the expected optimization time of the MMAS with construction graph $C(G)$ is bounded by $O(mn(\log n + \log w_{\max}))$.*

Proof. Let $e_1, \ldots, e_{n-1}$ be the edges of T^* and suppose w.l.o.g. that the edges of T are $e_1, \ldots, e_{n-2}, e'$ where $e' \neq e_i$ for $1 \leq i \leq n-1$. With probability $\Omega(1)$, exactly $n - 2$ (but not $n - 1$) out of the $n - 1$ nodes visited by the MMAS in $C(G)$ form a uniformly random subset of $\{e_1, \ldots, e_{n-1}\}$. Hence, e_{n-1} is missing with probability $1/(n - 1)$. Furthermore, the probability of visiting e' rather than e_{n-1} as the missing node has probability at least $\Omega(1/m)$. Hence, in total, T is constructed with probability $\Omega(1/(nm))$. Again we use the proof idea of Theorem 5.1. It suffices to show the following claim. Suppose the MMAS has constructed the spanning tree T^* in the last accepted solution. Let $T = T^* \setminus \{e\} \cup \{e'\}$ be any spanning tree that is obtained from T^* by including one edge e' and removing another edge e. Then the probability of producing T by the next constructed solution is $\Omega(1/(nm))$.

Let $e_1, \ldots, e_{n-1}$ be the edges of T^* and suppose w.l.o.g. that the edges of T are $e_1, \ldots, e_{n-2}, e'$ where $e' \neq e_i$ for $1 \leq i \leq n - 1$. We show that with probability $\Omega(1)$, exactly $n-2$ (but not $n-1$) out of the $n-1$ nodes visited by the MMAS in $C(G)$ form a uniformly random subset of $\{e_1, \ldots, e_{n-1}\}$. Hence, e_{n-1} is missing with probability $1/(n - 1)$. Furthermore, we show that the probability of visiting e' rather than e_{n-1} as the missing node has probability at least $\Omega(1/m)$. Hence, in total, T is constructed with probability $\Omega(1/(nm))$.

We still have to prove the statements on the probabilities in detail. We study the events E_i, $1 \leq i \leq n-1$, defined as follows. E_i occurs iff the first $i-1$ and the last $n - i - 1$ nodes visited by the MMAS (excluding s) correspond to edges of T^* whereas the ith one does not. Edges in $C(G)$ pointing to nodes of T^* have pheromone value h and all remaining edges have value ℓ. Hence, if $j - 1$ edges of T^* have been found, the probability of not choosing another edge of T^* by the next node visited in $C(G)$ is at most

$$\frac{(m - (n - 1))\ell}{((n - 1) - (j - 1))h} = \frac{1}{(n - j)\log n}.$$

Therefore, the first $i-1$ and last $n-i-1$ nodes (excluding s) visited correspond to edges of T^* with probability at least

$$1 - \sum_{j=1}^{n-2} \frac{1}{(n-j)\log n} \geq 1 - \frac{(\ln(n-1)+1)}{\log n} + \frac{1}{\log n} \geq 1 - \frac{\ln n}{\log n} = \Omega(1)$$

(estimating the $(n-1)$-th Harmonic number by $\ln(n-1)+1$) and, due to the symmetry of the update scheme, each subset of T^* of size $n-2$ is equally likely, i.e., has probability $\Omega(1/n)$. Additionally, the probability of choosing by the ith visited node an edge e' not contained in T^* equals

$$\frac{\ell}{(n-i)h + k\ell} \geq \frac{1}{(n-i+1)(m-n+1)\log n},$$

where k is the number of edges outside T^* that can still be chosen; note that $k\ell \leq h$. Hence, with probability at least $c/((n-i+1)mn\log n)$ for some small enough constant c (and large enough n), E_i occurs and the tree T is constructed. Since the E_i are mutually disjoint events, T is constructed instead of T^* with probability at least

$$\sum_{i=1}^{n-1} \frac{c}{(n-i+1)mn\log n} = \Omega(1/(mn))$$

as suggested. □

In the following, we examine the use of heuristic information for the Kruskal-based construction graph. Here it can be proven that strong heuristic information helps the MMAS mimic the greedy algorithm by Kruskal.

Theorem 5.15. *Choosing $\alpha = 0$ and $\beta \geq 6w_{\max}\log n$, the expected optimization time of the MMAS using the construction graph $C(G)$ is constant.*

Proof. We show that the next solution that the MMAS constructs is with probability at least $1/e$ a minimum spanning tree, where e is Euler's number. This implies that the expected number of solutions that have to be constructed until a minimum spanning tree has been computed is bounded above by e.

Let $(w_1, w_2, \ldots, w_{n-1})$ be the weights of edges of a minimum spanning tree. Let $w_i \leq w_{i+1}$, $1 \leq i \leq n-2$, and assume that the ant has already included $i-1$ edges that have weights $w_1, \ldots, w_{i-1}$, and consider the probability of choosing an edge of weight w_i in the next step. Let $M = \{e_1, \ldots, e_r\}$ be the set of edges that can be included without creating a cycle and denote by $M_i = \{e_1, \ldots, e_s\}$ the subset of M that includes all edges of weight w_i. W. l. o. g., we assume $w(e_i) \leq w(e_{i+1})$, $1 \leq i \leq r-1$.

The probability of choosing an edge of M_i in the next step is given by

$$\frac{\sum_{k=1}^{s}(\eta(e_k))^\beta}{\sum_{l=1}^{r}(\eta(e_l))^\beta} = \frac{\sum_{k=1}^{s}(\eta(e_k))^\beta}{\sum_{l=1}^{s}(\eta(e_l))^\beta + \sum_{l=s+1}^{r}(\eta(e_l))^\beta},$$

where $\eta(e_j) = 1/w(e_j)$ holds. Let $a = \sum_{k=1}^{s}(\eta(e_i))^\beta = \sum_{k=1}^{s}(1/w_i)^\beta$ and $b = \sum_{l=s+1}^{r}(\eta(e_l))^\beta$. The probability of choosing an edge of weight w_i is

$a/(a+b)$, which is at least $1 - 1/n$ if $b \leq a/n$. The number of edges in $M \setminus M_i$ is bounded above by m, and the weight of such an edge is at least $w_i + 1$. Hence, $b \leq m \cdot (1/(w_i + 1))^\beta$.

We would like $m \cdot (1/(w_i + 1))^\beta \leq s \cdot (1/w_i)^\beta / n$ to hold. This can be achieved by choosing

$$\beta \geq \frac{\log(mn/s)}{\log((w_i + 1)/w_i)} = \frac{\log(mn/s)}{\log(1 + 1/w_i)},$$

which is at most

$$(\log(mn/s))/(w_i/2) \leq 6w_{\max} \log n$$

since $mn \leq n^3$ and $e^x \leq 1 + 2x$ for $0 \leq x \leq 1$. Due to our choices, the ant traverses the edge with weight w_i with probability at least $1 - 1/n$. Therefore, the probability that in every step i such an edge is taken is at least $(1 - 1/n)^{n-1} \geq 1/e$, as suggested. □

The result of Theorem 5.15 does not necessarily improve upon Kruskal's algorithm since the computational efforts in a run of the construction algorithm and for initializing suitable random number generators (both of which are assumed constant in our cost measure for the optimization time) must not be neglected. With a careful implementation of the MMAS, however, the expected computational effort with respect to the well-known uniform cost measure could be at least bounded above by the runtime $O(m \log m)$ of Kruskal's algorithm.

Conclusions

The minimum spanning tree problem is one of the fundamental problems that is efficiently solvable. Several important variants of this problem are difficult, and stochastic search algorithms have a good chance of being competitive on them. As a first step towards the analysis of stochastic search algorithms on these problems, simple algorithms have been analyzed on the basic minimum spanning tree problem. The asymptotic worst-case (with respect to the problem instance) expected optimization time for simple evolutionary algorithms has been obtained exactly. The analysis is based on the investigation of the expected multiplicative distance decrease (with respect to the difference in the weights of the current graph and of a minimum spanning tree). The results presented in this chapter may be generalized to the computation of a minimum weight basis of a weighted matroid (Reichel and Skutella, 2007). On the other hand, it has been investigated for which search algorithms and problems one may consider a transformation of the weights such that better bounds can be obtained (Reichel and Skutella, 2009). This leads to an upper bound of $O(m^2 \log n)$ for $\mathrm{RLS}_b^{1,2}$ on f and f'. The bound for $(1+1)$ EA_b

given in this paper is $O(m^3 \log n)$, which is worse than the one presented in this chapter unless $w_{\max}$ is very large.

We studied simple ACO algorithms and investigated the impact of different construction graphs. In the case of the Broder-based construction procedure a polynomial, but relatively large, upper bound was proven. In addition, it was shown that heuristic information can mislead the algorithm such that an optimal solution is not found within a polynomial number of steps with high probability. In the case of the Kruskal-based construction procedure, the upper bound obtained shows that this construction graph leads to a better optimization process than simple evolutionary algorithms. In addition, a large influence of heuristic information makes the algorithm mimic Kruskal's algorithm for the minimum spanning tree problem.

6

Maximum Matchings

The maximum matching problem is a very well-studied combinatorial optimization problem. Given an undirected graph $G = (V, E)$, a matching is a
subset $E' \subset E$ of the edge set such that no two edges in E' share a common
endpoint. The maximum matching problem asks for a matching of maximum
cardinality. Such problems arise, e.g., in team planning when edges of a graph
denote possible collaborations of workers and the aim is to find a biggest partition of the workers into teams of size 2. Therefore, matching problems have
numerous generalizations to hypergraphs and weighted graphs, which will not
be discussed in this chapter. The maximum matching problem should not
be confused with the maximal matching problem, where the aim is to find a
subset of edges which is maximum with respect to inclusion, i.e., no proper
superset of the matching is a matching.

The maximum matching problem is solvable in polynomial time. The best
algorithms for the general case run in time $O(\sqrt{|V|} \cdot |E|)$ (Micali and Vazirani,
1980), which is also the best bound known for the special case of bipartite
graphs (Hopcroft and Karp, 1973). However, the algorithms for the latter
case are much simpler to describe and to analyze. All of them rely on the
fundamental concept of so-called *augmenting paths*, which will be explained
in detail below. Augmenting paths represent a way to improve the size of a
matching by performing local changes along the path. Hence, there is some
hope that locally searching algorithms are able to find maximum matchings.
This motivated early analyses of stochastic search algorithms for this problem,
most notably a study by Sasakik and Hajek (1988) with respect to simulated
annealing.

The contents of this chapter are based on and follow closely the works
by Giel and Wegener (2003, 2004, 2006), who concentrate on variants of randomized local search and (1+1) EA$_b$ for the maximum matching problem. In
Section 6.1, we describe the investigated search algorithms and fitness functions precisely and supply additional concepts for the analysis. Section 6.2
deals with a general result on the approximation capability of the algorithms.
For certain graph classes, exact solutions to the problem can be found in

F. Neumann, C. Witt, *Bioinspired Computation*
in Combinatorial Optimization, Natural Computing Series,
DOI 10.1007/978-3-642-16544-3_6, © Springer-Verlag Berlin Heidelberg 2010

expected polynomial time, which will be presented in Section 6.3. However, there are also graph instances for which no optimal solutions are found within polynomial time, with high probability. A worst-case result in this vein is described in Section 6.4.

6.1 Representations and Underlying Concepts

Giel and Wegener (2003, 2004, 2006) work with the following model of the maximum matching problem. Let $n = |V|$ denote the number of vertices and $m := |E|$ the number of edges of the graph for which a maximum matching is sought. Again, the encoding for binary search spaces is straightforward. When working with bitstrings of length m (!), a search point $s = (s_1, \ldots, s_m) \in \{0, 1\}^m$ is interpreted as the characteristic vector of the chosen subset of edges. If s describes a valid matching, the fitness function $f \colon \{0, 1\}^m$ just returns the number of chosen edges, i.e., $f(s) = s_1 + \cdots + s_m$. As in Chapter 5, several ways to handle invalid search points, in this case non-matchings, make sense. One way would be to assign large negative f values to them and to force the search algorithm to start from the empty matching, i.e., the all-zeros string. This kind of initialization is used in Sasaki's (1988) definition of simulated annealing. We stick with the uniform initialization used in the common stochastic search algorithms and introduce a component in the fitness function to direct the search towards valid matchings. The following idea is similar to that in Chapter 5 with the MST problem, where the number of connected components was to be minimized first in order to direct the algorithm to trees.

If $d(v) > 1$ edges incident on a vertex v are chosen by the search point s, a penalty $p(v)$ of value $d(v) - 1$ is assigned to the vertex; otherwise, $p(v) = 0$. Hence, exactly the vertices that are in accordance with the definition of a matching have no penalty. The penalty $p(s)$ of the search point is simply the sum of all vertex penalties, and the fitness function equals

$$f(s) = (-p(s), |s|_1).$$

This function has to be *maximized* in lexicographic order. As soon as a value of 0 has been obtained with respect to the first component, only valid matchings are considered.

The stochastic search algorithms studied are a randomized local search algorithm on the one hand and (1+1) EA$_b$ on the other hand. As for the MST problem studied in Chapter 5, a neighborhood of size 1 is not sufficient. In order for the search algorithm to accept a different matching, it can be necessary to flip out one edge and to include another edge. Of course, if there is an edge that is neither included in the current matching nor incident on another matching edge, then this edge can be chosen and leads to a larger matching. We call such edges *free* edges since both of its vertices are free, i.e., not incident on any matching edges. However, the lack of free edges does not mean that a matching is maximum. Instead, a characterization of optimality is based on the above-mentioned augmenting paths. We call a path through the vertices $v_1, \ldots, v_k$ an *alternating path* of length $k - 1$ with respect to a current matching if the edges $\{v_{2i}, v_{2i+1}\}$, $1 \leq i \leq k/2$, belong to the matching and the other edges do not belong to it. If additionally v_1 and v_k are free vertices, which is only possible for even k, i.e., an odd number of edges, then the path is called *augmenting*. In this case, we swap matching and non-matching edges along the augmenting path, which means that we remove from the matching all edges on the path that so far belong to the matching and add to the matching those edges on the path that do not belong to it. This procedure leads to a valid matching of increased (that is, "augmented") cardinality. A single free edge appears as the special case of an augmenting path of length 1. The following theorem is a fundamental characterization of optimal matchings.

Theorem 6.1 (Hopcroft and Karp, 1973). *A matching is of maximum cardinality if and only if there exists no augmenting path with respect to the matching.*

(1+1) EA_b can flip all edges of an augmenting path at once. A local search algorithm cannot do this in a single step. However, it can approach the improved matching by flipping two adjacent edges. If $v_1, \ldots, v_k$ is augmenting then $\{v_1, v_2\}$ can be turned into a matching edge and $\{v_2, v_3\}$ into a non-matching edge. This results in v_3 becoming free and $v_3, \ldots, v_k$ forming a new and shorter augmenting path. This motivates us to study the algorithm $\text{RLS}_b^{1,2}$ defined in Definition 2. The only change is that we apply both search algorithms for maximization, i.e., the condition $f(s') \leq f(s)$ is replaced by $f(s') \geq f(s)$ in the definition of both $\text{RLS}_b^{1,2}$ and (1+1) EA_b (see Algorithms 2 and 3).

6.2 Approximation Quality for General Graphs

The stochastic search algorithms start from a completely random string. The definition of the fitness function and the elitist selection of (1+1) EA_b and $\text{RLS}_b^{1,2}$ ensure that only matchings are accepted as future search points once a valid matching has been found. This happens efficiently as the following lemma shows.

Lemma 6.2. $RLS_b^{1,2}$ *and (1+1)* EA_b *find search points that represent matchings in expected time* $O(m \log m)$. $\square$

Proof. We argue as in the proof of the coupon collector's problem (Section 4.2.2). Let $k = -p(s)$ be the sum of the vertex penalties with respect to the search point s. Then k is less than $2m$, the sum of all vertex degrees.

Until a valid matching is found, only the first component of the fitness function is relevant, i.e., new search points are only accepted if they have a lower total penalty. By definition, there are at least $\lceil k/2 \rceil \leq m$ edges chosen by s, whose elimination decreases k. The probability of a specific 1-bit mutation equals $\Theta(1/m)$ for both algorithms. Hence, the expected waiting time to decrease k is bounded by $O(m/k)$. Summing up for $1 \leq k < 2m$ and estimating the Harmonic series according to $\sum_{k=1}^{2m} 1/k = O(\ln m)$ yields the claim.   □

The aim of this section is to show that the search algorithms are able to find good approximate solutions to the maximum matching problem for arbitrary graphs. The result is also based on the result by Hopcroft and Karp (1973). The main idea is as follows. Given a matching that is far away from optimality, there must not only be one, but many augmenting paths. The pigeonhole principle guarantees the existence of a relatively short augmenting path. This is made precise by the following lemma.

Lemma 6.3. *Let $G = (V, E)$ be a graph, M a non-maximum matching, and M^* a maximum matching. Then there exists an augmenting path with respect to M whose length is bounded from above by $L := 2\lfloor |M|/(|M^*| - |M|) \rfloor + 1$.*

Proof. Let $G' = (V, E')$ be the graph whose edge set is defined by $E' := M \oplus M^*$, where $\otimes$ denotes the symmetric difference, i.e., the exclusive OR of the search points. The graph G' consists of vertex-disjoint cycles and paths. Each cycle and each path of even length has the same number of M and M^* edges. Paths of odd length alternate between M and M^* edges. There is no odd-length path starting and ending with an M edge. Otherwise, it would be an augmenting path with respect to M^*. Hence, there are $|M^*| - |M|$ disjoint augmenting paths with respect to M. At least one has at most $\lfloor |M|/(|M^*| - |M|) \rfloor$ M edges and, therefore, at most L edges.   □

By means of the preceding lemma, we arrive at the announced result on the approximation quality of the search algorithms.

Theorem 6.4. *For $\epsilon > 0$, $RLS_b^{1,2}$ and (1+1) EA_b find a $(1 + \epsilon)$ optimal matching in expected time $O(m^{2\lceil 1/\epsilon \rceil})$ independently of the choice of the first search point.*

Proof. The first phase of the search finishes when a matching is found. By Lemma 6.3, this phase is short enough to be captured by the proposed runtime bound. Afterwards, let M be the current matching, and let M^* be an arbitrary maximum matching. The search is successful if $|M^*| \leq (1+\epsilon)|M|$. Otherwise, by Lemma 2, there exists an augmenting path for M whose length is bounded from above by $L := 2\lfloor |M|/(|M^*| - |M|) \rfloor + 1$. Since $|M^*| > (1 + \epsilon)|M|$, we conclude that

$$\frac{|M|}{|M^*| - |M|} < \epsilon^{-1}.$$

Consequently,

$$\left\lfloor \frac{|M|}{|M^*| - |M|} \right\rfloor \leq \begin{cases} \lfloor \epsilon^{-1} \rfloor = \lceil \epsilon^{-1} \rceil - 1 & \text{if } \epsilon^{-1} \text{ is not an integer,} \\ \lfloor \epsilon^{-1} \rfloor - 1 = \lceil \epsilon^{-1} \rceil - 1 & \text{if } \epsilon^{-1} \text{ is an integer.} \end{cases}$$

In any case, $L \leq 2\lceil 1/\epsilon \rceil - 1$.

The probability that (1+1) EA_b flips exactly the edges of an augmenting path of length ℓ is $(1/m)^\ell (1-1/m)^{m-\ell} = \Theta(m^{-\ell})$. The expected waiting time is therefore $\Theta(m^\ell)$. It is sufficient to wait $|M^*| \leq m$ times for such an event, where ℓ is always at most L. This proves the result for (1+1) EA_b.

$\text{RLS}_b^{1,2}$ can flip the augmenting path in $\lfloor \ell/2 \rfloor + 1$ steps. In each of the first $\ell/2$ steps, the length of the augmenting path is decreased by 2 by flipping the first two or the last two edges, and in the last step the remaining edge of the augmenting path is flipped. The probability that a phase of length $\lfloor \ell/2 \rfloor + 1$ is successful is bounded from below by $\Omega((m^{-2})^{\lfloor \ell/2 \rfloor} \cdot m^{-1}) = \Omega(m^{-\ell})$, where we used the fact that the length ℓ of an augmenting path is odd. The expected number of unsuccessful phases preceding a successful phase is $O(m^\ell)$. Again, we have $\ell \leq L$. The difference with the case of (1+1) EA_b is that a phase may consist of more than one step. However, in each step the probability that a phase is continued successfully is bounded from above by $O(m^{-1})$. Hence, the expected phase length is $O(1)$. This also holds under the assumption that a phase is unsuccessful. The length of the successful phase equals $\lfloor \ell/2 \rfloor + 1$. Hence, the expected number of steps to improve the matching again is bounded from above by $O(\ell + m^\ell) = O(m^\ell)$, which proves the theorem. $\square$

The previous theorem also implies that the simple stochastic search algorithms are PRASs (polynomial-time randomized approximation schemes) in the sense of Definition 2.8. The following corollary, which follows from Theorem 6.4 by using Markov's inequality, makes this explicit. We just let c be a constant such that Theorem 6.4 holds for the bound $c \cdot m^{2\lceil 1/\epsilon \rceil}$.

Corollary 6.5. *If we run $RLS_b^{1,2}$ or (1+1) EA_b for $4cm^{2\lceil 1/\epsilon \rceil}$ iterations, we obtain a PRAS for the maximum matching problem, i.e., independently of the choice of the first search point, the probability of producing a $(1 + \epsilon)$ optimal solution is at least $3/4$.*

6.3 Upper Bounds for Simple Graph Classes

After seeing that $\text{RLS}_b^{1,2}$ and (1+1) EA_b find good approximations to maximum matchings in expected polynomial time, we are interested in graphs where even maximum matchings can be found in expected polynomial time. We start with the simple graph called *path*. As the name suggests, it consists of a path of m edges. This graph allows a matching of maximal size for connected graphs, namely $\lceil m/2 \rceil$. The analysis of the search algorithms on this graph contains typical aspects of their behavior on more complicated instances.

As in Chapter 5, there are typically many steps of $\mathrm{RLS}_b^{1,2}$ and $(1+1)$ EA_b leading to infeasible search points, in this case search points that do not encode a matching. Such steps slow the search algorithms down, but cannot be avoided without introducing problem-specific knowledge, which is not always available. In the forthcoming analysis, we account for such steps by the consideration of so-called *relevant* steps, where the exact definition will depend on the situation. Denoting by R the number of relevant steps and by T the total number of steps until a certain goal is achieved, the following argumentation is typical. If an expected number of $E(R)$ relevant steps is necessary and every step is relevant with probability at least p, then the expected total number of steps $E(T)$ is at most $p^{-1} \cdot E(R)$.

We start with a simple upper bound for $\mathrm{RLS}_b^{1,2}$ on the path graph.

Fig. 6.1. An augmenting path (indicated by gray area) and environments of possible mutations leading to extensions or shortenings (dotted)

Theorem 6.6. *For a path of m edges, the expected optimization time of $RLS_b^{1,2}$ is $O(m^4)$ independently of the choice of the first search point.*

Proof. By Lemma 6.2, the expected waiting time for a matching is small enough to be captured by the $O(m^4)$ bound. The size of a maximum matching equals $\lceil m/2 \rceil$. If the current matching size is $\lceil m/2 \rceil - i$, there exist at least i augmenting paths and one of length at most $\ell := m/i$. In every step, we conceptually select a shortest augmenting path P; hence the considered P might be different in the course of optimization. Now a step is called Prelevant if it is accepted and P is altered. The probability of a Prelevant step is $\Omega(1/m^2)$. This is due to the following observations. If the length of P is at least 3, it is lower bounded by the probability that a pair of edges at one end of P flips; otherwise the path consists of only a free edge, and the considered probability is even $\Omega(1/m)$. If we can show that an expected number of $O(\ell^2)$ Prelevant steps is sufficient to improve the matching by one edge, then $\sum_{i=1}^{\lceil m/2 \rceil} O((m/i)^2) = O(m^2)$ Prelevant steps are sufficient, and the expected optimization time is $O(m^4)$.

If $|P| \geq 3$, there are no free edges. Only mixed mutation steps, where a non-matching edge and a matching edge flip, can be accepted. Since each non-matching edge e has at least one neighbor e' in the matching, e' must flip, too. That means that only a non-matching edge e incident on a free vertex together with a matching edge e' such that e and e' have an endpoint in common can flip. In the considered case of a path graph, only pairs of neighbored edges located at one end of an alternating path can flip in accepted steps. Such a pair consists of either two neighbored edges outside P or two edges inside P, in both cases with one edge incident on an endpoint of P. (See Figure 6.1

for an illustration, where either the first two or the last two edges in one of the dotted boxes are allowed to flip together.) The first case increases the length of P and the second one decreases it. Since P might be aligned with an endpoint of the whole graph itself, the situation can even be in favor of decreasing steps. Hence, P is shortened with probability at least $1/2$ if a pair of neighbored edges flips. If $|P| = 1$, the probability that the length of P is decreased to 0 is at least $1/(2m)$ since a 1-bit mutation of $\mathrm{RLS}_{\mathrm{b}}^{1,2}$ is sufficient. In contrast, the probability that the path grows at either end is at most $2 \cdot (1/2) \cdot \binom{m}{2}^{-1} = 2/(m(m-1))$ in this case. Hence, the conditional probability that the next Prelevant step is decreasing is at least $1/(1 + 4/(m - 1)) \geq 1/2$, for $m \geq 5$.

Taking the two cases together, we are confronted with a random walk on the numbers $\{0, 1, 3, 5, \ldots, \ell\}$ describing the current length of P. This walk goes from a state to the lower neighboring state with probability at least $1/2$ and to the higher neighboring state otherwise. Since we are interested in reaching state 0, we may pessimistically assume the transition probabilities to be exactly $1/2$ and arrive at the scenario relevant for Theorem 4.7. The graph on which the random walk takes place is itself a path; hence its number of edges is trivially bounded by ℓ. The time to reach state 0, i.e., one end of this path, is bounded by the cover time for the graphs, which is $O(\ell^2)$ using Theorem 4.7. □

Basically the same ideas as in the proof of Theorem 6.6 can be used to prove also the bound $O(m^4)$ for (1+1) EA_{b}. However, the analysis is complicated by the fact that the latter search algorithm can flip many bits in a step. We are only interested in Prelevant steps. For our analysis, we define Pclean steps, which are Prelevant steps causing only small changes in P. Then, a phase including $\Theta(\ell^2)$ Prelevant steps is called Pclean if all its Prelevant steps are Pclean. The idea is to prove that a phase is Pclean with probability $\Omega(1)$ and that a Pclean phase plus the next Prelevant step improve the matching with probability $\Omega(1)$.

Theorem 6.7. *For a path of m edges, the expected optimization time of (1+1) EA_b is $O(m^4)$ independently of the choice of the first search point.*

Fig. 6.2. An augmenting path (indicated by gray area) and the environments E_u and E_v (dotted); free vertices are indicated by a circle

Proof. For the definition of P, ℓ, and Prelevant steps, see the proof of Theorem 6.6. With the same arguments as used there, it suffices to prove that the expected number of Prelevant steps to improve the matching is $O(\ell^2)$.

Pclean steps are only defined for situations without free edges. Let u and v be the endpoints of P, and let E_u be the set of edges where at least one endpoint has at most a distance of 3 from u, and analogously for E_v (see Figure 6.2 for an illustration). Then we call a Prelevant step a Pclean step if

- at most three edges in $E' := E_u \cup E_v$ flip and
- at most two of the flipping edges in E' are neighbors.

We describe the effect of clean steps on P. The free vertices partition the graph into alternating paths (see also Figure 6.2 for an example). As there is no free edge, there is an augmenting path of at least three edges between a free vertex and the next free vertex. Hence, a Pclean step cannot flip all edges of P because this would require flipping a block of three edges in E'. Consequently, P cannot vanish in a Pclean step; however, it is possible that new free vertices are created between u and v. Then, we interpret this event as a step shortening P by at least two edges. It is impossible that a Pclean step lengthens P by more than two edges, i.e., at least four edges, since this requires flipping more than three edges in E'. Thus, Pclean steps lengthen P only by 2, and to this end it is necessary to flip a pair of neighbored edges outside the augmenting path but touching either u or v (the situation already discussed in Figure 6.1). For a Pclean step to decrease the length of P by at least 2, it is sufficient to flip a pair of neighbored edges at either end of the augmenting path (see again Figure 6.1). Since at most three edges of E' may flip, at most one of the discussed pairs of neighbored edges can flip in a Prelevant step. Hence, Prelevant steps either lengthen or shorten P, and the probability of shortening steps is only larger than the probability of lengthening steps.

As the aim of a phase is to produce an improved matching or some free edge, it is convenient to include these good events into Pclean steps. We broaden our definition of Pclean steps and call accepted steps that produce a free edge or improve the matching Pclean, too. Now, we upper bound the probability of Prelevant but not Pclean steps (in situations without free edges). A necessary event to violate the first property is that four out of at most 16 edges of E' flip. The probability of this event is $O(1/m^4)$. For the second property, let k denote the length of the longest block B of flipping edges in E'. The probability that a block of length $k \geq 4$ flips is upper bounded by the probability of the event that one out of at most ten potential blocks of length 4 in E' flips. The probability of this event is $O(1/m^4)$. A mutation step where $k = 3$ produces a local surplus of either one non-matching edge or one matching edge in B. If the surplus is not balanced outside B, the step is either not accepted because the fitness would decrease or the step is clean because the matching is improved. To compensate for a surplus of one non-matching edge, one more non-matching edge than the number of flipping matching edges must flip elsewhere. This may be a non-matching edge next to B but outside E' if B is located at a border of E'. The probability of such a step is only $O(1/m^4)$. If B is not located at a border of E', another block B' of at least three edges not neighboring B has to flip. This results in a probability of at most $O((1/m^3) \cdot (m \cdot 1/m^3)) = O(1/m^5)$. If a local surplus of one matching edge has to be balanced, either only another matching edge flips and, because a free edge is created, the step is clean, or another block of at least three edges flips. The probability of the last possibility is again $O(1/m^5)$. Altogether, the probability of a Prelevant but not Pclean step is $O(1/m^4)$, and the conditional probability that a Prelevant step is not Pclean is $O(1/m^2)$. Hence, a phase of $O(\ell^2) = O(m^2)$ Prelevant steps is clean with probability $\Omega(1)$.

Pessimistically assuming that shortenings shorten the path by exactly two edges and that the probability of shortening in a clean, relevant step is exactly $1/2$, we treat this as a fair random walk as in the last paragraph of the proof of Theorem 6.6. Hence, an expected number of $O(\ell^2)$ clean relevant steps reduces the length of P to at most 1. By Markov's inequality, this happens with probability $\Omega(1)$ in $c\ell^2$ clean relevant steps if c is a large enough constant. Afterwards, at least one free edge exists, and a step is Prelevant with a probability of $\Omega(1/m)$, Hence, the next Prelevant step improves the matching with probability $\Omega(1)$. $\square$

The results from the previous two theorems deserve some discussion. On the one hand, paths are difficult since augmenting paths tend to be rather long in the final stages of optimization. On the other hand, paths are easy since there are not many ways to lengthen an augmenting path. As indicated above, the relatively large time bound $O(m^4) = O(n^4)$ can the explained by the characteristics of general (and somehow blind) local search. As the search algorithm does not "know" that only matchings are valid search points, it keeps wasting a lot of steps by producing invalid search points and rejecting them immediately. Moreover, while the analysis focuses on a shortest augmenting path, there may be many steps which alter the search point at a completely different place. In the case of $O(1)$ augmenting paths and no selectable edge, a step is relevant only with a probability of $\Theta(1/m^2)$, and the expected number of relevant steps is $O(m^2) = O(n^2)$. Actually, the search on the level of second-best matchings is responsible for this. If the number of edges is odd, the path graph has a unique maximum matching consisting of $\lceil m/2 \rceil$ edges. Therefore, any second-best matching of size $\lfloor m/2 \rfloor = \lceil m/2 \rceil - 1$ has only one augmenting path P. We show that the simple search algorithms have an expected optimization time of $\Omega(m^4)$ if the initial situation is a second-best matching and P is not too short.

Theorem 6.8. *For a path of m edges, m odd, the expected optimization time of $RLS_b^{1,2}$ and (1+1) EA_b is $\Theta(m^4)$ if the initial situation is a second-best matching with an augmenting path of length $\Omega(m)$.*

Proof. The upper bounds follow from Theorems 6.6 and 6.7. For the lower bounds, we would like to exploit the properties of the random walk describing the length of the augmenting path, analyzed in the two theorems and illustrated in Figure 6.1. Note that only 2-bit flips of $RLS_b^{1,2}$ are possible in relevant steps. Hence, as long as the augmenting path P is not adjacent to a border of the path graph itself and at most two edges flip, we are confronted with a fair random walk increasing or decreasing the length of the path with probability $1/2$ each in relevant steps. Only if P is at a border can the probability of decreasing the length be greater than $1/2$ in relevant steps. This corresponds to the scenario of the gambler's ruin theorem with $p = q = 1/2$ (see Theorem 4.8) except for the fact that the game might be changed when P touches a border. If we assume P to be at distance $\Omega(m)$ from both borders and to have initial length $\Omega(m)$, then an endpoint of the path has to move by a distance of $\Omega(m)$ or the whole path has to shrink in length by at least $\Omega(m)$ before the process differs from the fair gambler's ruin game. Using $a = \Omega(m)$ and $b - a = \Omega(m)$ in Theorem 4.8, the expected number of steps needed for the process to move by at least $\Omega(m)$ states is $\Omega(m^2)$. It is easy to see that $\Omega(m^2)$ is not only a lower bound on the expectation, but that $\Omega(m^2)$ relevant steps are also needed with probability $\Omega(1)$. (If the latter did not hold, we would immediately obtain a better bound on the expectation by repeating independent phases.) Since a relevant step has probability $\Theta(1/m^2)$, the lower bound for $\mathrm{RLS}_b^{1,2}$ follows.

For $(1+1)$ EA_b, the considered 2-bit flips have also probability $\Theta(1/m^2)$ but we must also take into account $2k$-bit flips for $k \geq 2$. We pessimistically assume that the latter only decrease the length of P and show that this additional decrease is at most half the initial length of P. Then, the length of P is always at least $\Omega(m)$ and the probability of a step flipping exactly the edges of P is small enough. If $2k$ edges flip in an accepted step, they form one or two blocks where the last or first edge of a block is adjacent to one of the exposed endpoints of P. Thus, there are $O(k)$ possibilities for an accepted $2k$-bit flip, and the expected decrease by means of $2k$-bit flips in a single step is $2k \cdot O(k/m^{2k}) = O(k^2/m^{2k})$. The sum for all $k \geq 2$ is $O(1/m^4)$. Hence, the expected decrease by steps flipping more than two bits is $O(1/m^4)$ in each step. Within βm^4 steps, this expected decrease is $O(1)$ and the decrease is less than half the initial length of P with probability $1 - o(1)$ if the constant $\beta > 0$ is small enough. $\square$

Giel and Wegener (2004, 2006) extend the previous results from paths to trees, i.e., connected graphs without cycles. They conjecture that path graphs represent the most difficult instance within the class of tree graphs since a path is a tree with maximal diameter and the diameter bounds the length of a longest augmenting path.

We do not present the complete involved analysis that Giel and Wegener (2004, 2006) perform for $\mathrm{RLS}_b^{1,2}$ on trees since such a presentation would be beyond the scope of this book. However, it is possible to present the general idea behind why $\mathrm{RLS}_b^{1,2}$ finds maximum matchings on complete trees in expected polynomial time. More precisely, the authors obtain the following theorem.

Theorem 6.9. *The expected time until* $RLS_b^{1,2}$ *finds a maximum matching on a complete $k$ary tree, $k \geq 2$, is bounded by $O(m^{7/2})$ independently of the choice of the first search point.*

When $RLS_b^{1,2}$ operates on complete $k$ary trees, there are two essential differences with respect to the path graph. Given a situation without free edges and an augmenting path P shorter than the diameter, there must be a free vertex v at one end of P that is not a leaf (vertex of degree 1) of the graph. This means that v must have degree k, which implies that there are $k-1$ ways to lengthen and only one way to shorten P. Each of these possibilities is chosen with the same probability, and for $k \geq 2$, $RLS_b^{1,2}$ is confronted with an unfair game that is biased towards increasing the length of P. In terms of the gambler's ruin theorem (Theorem 4.8), the event of P reaching its maximal length D, where D is the diameter of the graph, corresponds to the gambler's ruin. The probability of the gambler's gain, i.e., reaching length 0 before length D, starting from a worst-case length $D - 1$, equals

$$1 - \frac{(k-1)^D - (k-1)}{(k-1)^D - 1} = \frac{1}{(k-1)^{D-1}},$$

which is exponentially small in D. On the other hand, it holds that $D \leq 2\lceil \log_k m \rceil$ since the depth of the $k$ary tree is at most $\lceil \log_k m \rceil$. Inserting this into the above formula results in a probability of $\Omega(1/poly(n))$ for the gambler's gain. Since the expected length of the unfair game is also polynomial (Theorem 4.8), we obtain an overall expected polynomial time until the matching is improved.

Finally, using much more sophisticated arguments, Giel and Wegener (2004, 2006) extend the analysis to the case of arbitrary trees. They obtain the following theorem.

Theorem 6.10. *The expected time until* $RLS_b^{1,2}$ *finds a maximum matching in a tree with diameter D is bounded by $O(D^2 m^4)$ independently of the choice of the first search point.*

The authors also believe that basically the same results hold for $(1+1)$ EA_b, but in the case of arbitrary trees it is much more difficult to control the effect of steps flipping more than two bits than it is for the path graph. This concludes the presentation of the positive results. In the following section, we explore the limits of the search algorithms.

6.4 A Worst-Case Result

The result of Theorem 6.4 shows that $RLS_b^{1,2}$ and $(1+1)$ EA_b represent good approximation algorithms for the maximum matching problem. However, in the worst case they are not able to find an optimum in expected polynomial time. The analysis by Giel and Wegener (2003, 2006) is based on a graph class that was introduced by Sasakik and Hajek (1988). The graph $G_{h,\ell}$ for odd $\ell = 2\ell' + 1$ is best illustrated by placing its $n := h(\ell+1)$ vertices in h rows and $\ell + 1$ columns on a grid, i.e., $V = \{(i,j) \mid 1 \le i \le h, 0 \le j \le \ell\}$. Between column j, j even, and column $j + 1$, there are exactly the horizontal edges $\{(i,j),(i,j+1)\}$, $1 \le i \le h$. In contrast, there are complete bipartite graphs between column j and column $j + 1$ for odd values of j. The graph $G_{3,11}$ is shown in Figure 6.3. The unique perfect matching M^* consists of all horizontal edges between the columns j and $j+1$ for even j. The set of all other edges is denoted by $\overline{M}^*$. Obviously, we have $m = |M| + |\overline{M}^*| = (\ell'+1)h + \ell' h^2 = \Theta(\ell h^2)$ for the number of edges.

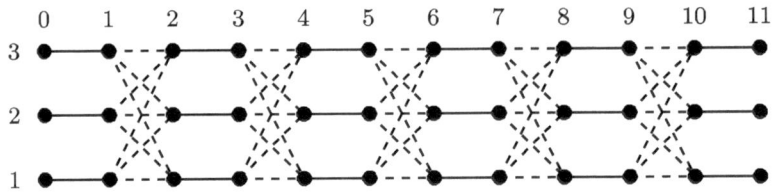

Fig. 6.3. The graph $G_{h,\ell}$, $h = 3$, $\ell = 11$, and its perfect matching

For the forthcoming analyses, it is sufficient to consider second-best (also called *almost perfect*) matchings of size $|M^*| - 1$ for the graph $G_{h,\ell}$ and to show that the final improvement takes in expectation an exponential time. Given an almost perfect matching, there is only one unique augmenting path left (a formal proof for this fact is already contained in the proof of Lemma 6.3). This augmenting path has the following properties.

Lemma 6.11. *Let Q be the unique augmenting path for an almost perfect matching in the graph $G_{h,\ell}$. Then*

- *Q "runs from left to right", i.e., it contains at most one vertex from each column,*
- *if the endpoints of Q are not in the first or last column, there are $2h$ lengthenings and two shortenings by 2-bit flips; otherwise there are h lengthenings and still two shortenings.*

Proof. For the first property, assume that two vertices belonging to the same column both lie on Q. Due to the structure of $G_{h,\ell}$, this implies in particular the existence of an odd column j with this property. Then Q runs along adjacent edges $e' = \{(i',j),(i,j+1)\}$ and $e'' = \{(i,j),(i,j+1)\}$, both of which are in $\overline{M}^*$. Either e' or e'' is contained in the almost perfect matching, and w. l. o. g., this is the case for e'. We consider Q as running from left to right in row i', then changing direction via e' and e'', and subsequently running to

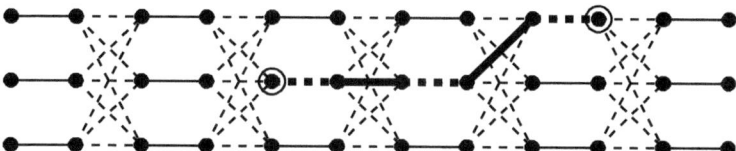

Fig. 6.4. An almost perfect matching in $G_{h,\ell}$ and its augmenting path; the free vertices are marked by a circle

the left in row i. We exploit the fact that e'' is not included in the almost perfect matching. Hence, as long as Q continues along row i to the left, the almost perfect matching can only contain M^* edges. This implies that Q will not change to another row again and ends at a free vertex (i, j') in the very same row. If also $(i, j' - 1)$ is a free vertex, then edge $\{(i, j' - 1), (i, j')\}$ is free; hence there is another augmenting path in contradiction to the fact that we have an almost perfect matching. If $(i, j' - 1)$ is not free, the almost perfect matching must include another $\overline{M}^*$ edge, which again implies the existence of another augmenting path.

As a consequence of the first property and the fact that we are dealing with an almost perfect matching, at least one endpoint of Q is adjacent to h different $\overline{M}^*$ edges that are not contained in the almost perfect matching. If the other endpoint of Q is not in the first or last column, this also holds for the other endpoint. This implies the second property. $\square$

Lemma 6.11 implies that the search algorithms are confronted with an unfair game if $h \geq 3$. The tendency towards increasing the length of Q will provably result in an exponential expected optimization time. To formalize this, we use a potential function, denoted by P, which maps search points (i.e., selections of edges) to integral values. For the sake of convenience, this function is defined only for almost perfect matchings and denotes the current length of the unique augmenting path for such a matching; hence, P takes only odd values. Note that P is not injective. In particular, we cannot determine from the Pvalue whether the augmenting path is adjacent to a border of the $G_{h,\ell}$. Assuming a worst-case perspective, we assume one endpoint of the path to be at a border. Then there are, according to Lemma 6.11, h 2-bit flips increasing and two such flips decreasing the potential. Giel and Wegener (2003, 2004) consider also the case $h = 2$, which is of some special interest since $G_{2,\ell}$ is the only planar graph in this class of graphs. However, much additional effort is required in this case to show that the augmenting path is likely to be far away from a border and that the game is, therefore, still unfair. In the following, we present the analysis only for the case $h \geq 3$.

The proof strategy for an exponential lower bound with respect to the search algorithms $\mathrm{RLS}_b^{1,2}$ and $(1+1)$ EA_b is as follows. As already mentioned, we consider only second-best matchings. Starting from such a search point, with overwhelming probability, $O(m^3)$ steps are enough to obtain either the perfect matching or an augmenting path of maximal length, which is ℓ according to Lemma 6.11. We estimate the probability of which of these two events happens first. If the augmenting path has reached length ℓ, we prove that it is very likely to need exponentially many steps to obtain the perfect matching. To obtain this result, it is required that ℓ be polynomial in m. We are mainly interested in the case $3 \le h \le \ell$, implying that $\ell = \Omega(m^{1/3})$. Then, 2^ℓ is exponential in m. In a phase of this length, it is quite likely that (1+1) EA performs certain steps of exponentially small probability, which can change the current matching at a significant number of places and not only locally. Therefore, the following analysis will again be easier to conduct for $RLS_b^{1,2}$.

We concentrate now on specific Pvalues. If the Pvalue is 1, i.e., we have a free edge, it is likely we will find the perfect matching in the next step.

Lemma 6.12. *For $RLS_b^{1,2}$ and (1+1) EA_b starting with an almost perfect matching with a Pvalue of 1, the following holds. The probability of reaching an almost perfect matching with a Pvalue of at least 3 is $\Theta(h/m)$.*

Proof. Since $P = 1$, the augmenting path consists of a free edge. To improve the matching, it is sufficient that only the free edge flip, and it is necessary that this edge flip. Therefore, the probability of creating the perfect matching is $\Theta(1/m)$. To increase P, it is sufficient that one of the h or $2h$ edge pairs lengthening the augmenting path flip. (If also the free edge flips, the path moves to another position. Then, at least a matching edge has to flip and additionally, one of the h or $2h$ pairs lengthening this augmenting path.) Hence, the unconditional probability of reaching a situation where $P \ge 3$ equals $\Theta(h/m^2)$. The conditional probability of reaching $P \ge 3$ rather than $P = 0$ is, therefore, $\Theta(h/m)$. $\square$

We prove the worst-case result for $RLS_b^{1,2}$ first. Assuming a Pvalue of at least 3, we will apply the results of the gambler's ruin theorem from Theorem 4.8. From the perspective of a lower bound, increasing the Pvalue to its maximum before improving the matching is a success. Hence, increasing P by 2 (recall that only odd values are taken) corresponds to the gambler's winning a unit of money and decreasing P corresponds to his losing a unit. The probability p_h of winning is at least $h/(h+2)$ if the Pvalue is at least 3, pessimistically assuming that one endpoint of the augmenting path is at a border of $G_{h,\ell}$. Since only $h \ge 3$ is considered, we have $p_h \ge 3/5$, i.e., an unfair game. More generally, we obtain $r_h = (1 - p_h)/p_h \le 2/h$ for the setup of Theorem 4.8. Given an initial Pvalue of $P_0 \ge 3$, the probability of reaching a Pvalue of ℓ before a value of at most 1 is at least

$$1 - \frac{r_h^{\ell'} - r_h^{P_0'}}{r_h^{\ell'} - 1} = \frac{r_h^{P_0'} - 1}{r_h^{\ell'} - 1} = \frac{1 - r_h^{P_0'}}{1 - r_h^{\ell'}},$$

where $\ell' := \lfloor \ell/2 \rfloor$ is the number of different values greater than 1 the Pvalue can take and $P_0' := \lfloor P_0/2 \rfloor$ is the number of possible Pvalues greater than 1 and less than or equal to P_0. Since $r_h < 1$, the probability under consideration is at least $1 - r_h^{P_0'}$. Our considerations are summarized by the following lemma, which pessimistically assumes the matching to be improved once the Pvalue has dropped to 1.

Lemma 6.13. *For $RLS_b^{1,2}$ starting with an almost perfect matching with a Pvalue of $P_0 \geq 3$, the probability of constructing an augmenting path of maximal length before the perfect matching is at least $1 - (2/h)^{\lfloor P_0/2 \rfloor}$.*

For (1+1) EA$_b$, we have to estimate the probabilities of steps where many flipping bits influence the augmenting path. In order to simplify the analysis, we interpret the following event as a loss of the whole game. At least the leftmost $i \leq 4$ and the rightmost $j \geq 4 - i$ edges of the augmenting path flip. The probability of this event is bounded from above by $O(1/m^4)$. Now, the only way of decreasing the Pvalue by 1 without losing the game is by flipping exactly the two leftmost or the two rightmost edges of the augmenting path. The probability of this event equals $2(1/m)^2(1-1/m)^{m-2}$. This leads basically to the same probabilities as in Lemma 6.13, but we have to take into account the probability of $\Theta(1/m^3)$ of turning a search point with a short augmenting path of length 3 into the perfect matching. We obtain the following result, which provides in essence the same bounds as the preceding lemma.

Lemma 6.14. *For (1+1) EA$_b$ starting with an almost perfect matching with a Pvalue of $P_0 \geq 3$, the probability of constructing an augmenting path of maximal length before the perfect matching is at least $1 - O(1/m) - ((2/h) + O(1/m))^{\lfloor P_0/2 \rfloor}$.*

Proof. Since we pessimistically consider the event of a Pvalue of 1 as the event that the perfect matching is created, we can include the event that an augmenting path of length 3 is flipped in the event in which the gambler loses one unit of money. Since 2-bits are necessary and sufficient, the probability that a step changes the augmenting path is $\Theta(1/m^2)$. The probability of flipping an augmenting path of length 3 is $\Theta(1/m^3)$. Therefore, it is sufficient to increase the value of r_h from the above analysis by $O(1/m)$.

In addition, there is a probability of $O(1/m^4)$ for each step that the game is immediately lost because the Pvalue changes by more than 1. If we can prove that the game is finished anyway within $O(m^3)$ steps with probability at least $1 - O(1/m)$, the probability of observing a step of probability of $O(1/m^4)$ until the end of the game is $O(1/m)$. This is accounted for by first the term $-O(1/m)$ in the bound of the lemma.

We are left with the claim on the number of steps until the end of the game. Using $p_h > 1/2$ in Theorem 4.8, we obtain $D_{\lfloor P_0/2 \rfloor} = O(\ell)$, which means that the expected number of steps of (1+1) EA$_b$ is $O(\ell m^2) = O(m^3)$ since a step is relevant, i.e., changes the length of the path, with probability $\Omega(m^2)$. The bound $O(m^3)$ holds also with probability $1 - 2^{-\Omega(m)}$. By Chernoff bounds, there is with probability $1 - 2^{-\Omega(m)}$ a surplus of at least $\lfloor \ell/2 \rfloor$ increasing steps within some cm relevant steps, c an appropriate constant. Also by Chernoff bounds, there are with probability $1 - 2^{-\Omega(m)}$ at least cm relevant steps within $c'm^3$ steps of $(1+1)$ EA$_b$, c' another appropriate constant. This proves the lemma. □

Putting the previous arguments presented until Lemma 6.13 together, we obtain a first lower bound on the runtime of RLS$_b^{1,2}$.

Lemma 6.15. *Starting with an almost perfect matching and an augmenting path of maximal length, the probability that RLS$_b^{1,2}$ finds the perfect matching within $2^{c\ell}$ steps, $c > 0$ an appropriate constant, is bounded from above by $2^{-\Omega(\ell)}$.*

Proof. We essentially apply the argumentation leading to Lemma 6.13. Starting from a Pvalue of ℓ, the value $k := \lceil \ell/2 \rceil = \Omega(\ell)$ or, if k is even, the value $k - 1$ has to be taken at least once before 0 is reached. Starting from $k - 1$ (analogously for k), we have an unfair game where the probability of the gamber not winning (where winning means returning to a Pvalue of ℓ) is bounded from above by

$$\frac{1 - r_h^{\lfloor (k-1)/2 \rfloor}}{1 - r_h^{\lfloor \ell/2 \rfloor}} \leq 2^{-c'\ell}$$

for some constant $c' > 0$. The game is repeated until it is lost (i.e., a Pvalue of at most 1 is reached) for the first time. The probability of losing at least once in $2^{c\ell}$ games is bounded from above by $2^{(c-c')\ell} = 2^{-\Omega(\ell)}$ if the constant $c < c'$ is chosen small enough. □

To prove a corresponding result for $(1+1)$ EA$_b$, we still exploit the fact that the game is unfair, i.e., there is a drift towards increasing the Pvalue. However, we cannot apply the gambler's ruin theorem any longer since, as mentioned above, exponentially long phases allow for steps that change the situation in a significant number of places. Therefore, the simplified drift theorem (Theorem 4.9) will be applied.

Lemma 6.16. *Starting with an almost perfect matching and an augmenting path of maximal length, the probability that $(1+1)$ EA$_b$ finds the perfect matching within $2^{c\ell}$ steps, $c > 0$ an appropriate constant, is bounded from above by $2^{-\Omega(\ell)}$.*

Proof. To apply Theorem 4.9, we set $a := 0$ and $b := \lceil \ell/2 \rceil - 1$. The random variables X_t are obtained by taking the random Pvalues at the respective time points, dividing them by 2 and rounding the result up. In this way, we obtain a process on the state space $\{0, 1, \ldots, \lceil \ell/2 \rceil\}$.

Given a current $X_t$value of i, where $i \leq \lceil \ell/2 \rceil - 1$, we need an estimate of the expected change of this value. The probability of increasing the value by 1, i.e., lengthening the augmenting path of length $2i - 1$ by 2, is bounded from below by

$$p_1(i) \geq \frac{h}{m^2} \left(1 - \frac{1}{m}\right)^{m-2}$$

since at least one end of the path is not at a border of $G_{h,\ell}$ and there are h appropriate 2-bit flips. Here we use the fact that $i \leq \lceil \ell/2 \rceil - 1$, i.e., the augmenting path can still be lengthened. On the other hand, the probability of decreasing the $X_t$value by $j \geq 2$ is bounded from above according to

$$p_{-j}(i) \leq (j+1) \cdot \left(\frac{1}{m}\right)^{2j}$$

since it is necessary to flip the $2k$ leftmost edges and the $2(j - k)$ rightmost edges of the augmenting path for some $k \in \{0, \ldots, j\}$. For $p_{-1}(i)$, we need a better bound that is at least by a constant factor smaller than $p_1(i)$. We estimate

$$p_{-1}(i) \leq \frac{2}{m^2} \left(1 - \frac{1}{m}\right)^{m-2} + \frac{3}{m^4}$$

since there are exactly two ways of flipping exactly two edges, and otherwise one has to flip at least the $2k$, $0 \leq k \leq 2$, leftmost edges and the $4 - 2k$ rightmost edges of the augmenting path.

Since most other mutations of $(1+1)$ EA$_b$ will be rejected in this setting due to worse fitness, we use the condition C_{rel} that a step is *relevant*, meaning it is accepted and changes the matching. Of course, if we obtain a lower bound on the required number of relevant steps, this also bounds the actual number of steps of $(1+1)$ EA$_b$ from below. The probability p_{rel} of a relevant step is bounded according to

$$\frac{1}{m^2} \left(1 - \frac{1}{m}\right)^{m-2} \leq p_{\mathrm{rel}} \leq \frac{2h+2}{m^2}.$$

The lower bound holds because, unless the optimum has been found, there always are two edges that, when flipped, lengthen or shorten the augmenting path. The upper bound holds because there are at most $2(h + 1) = 2h + 2$ couples of edges adjacent to a border of the augmenting path that, when flipped, lengthen or shorten the path. The probability that more than two bits flip and the step is relevant is lower since at least one of the $2h + 2$ couples considered in the bound has to be flipped anyway.

Let $R(i) = (\Delta(i) \mid C_{\mathrm{rel}})$ denote the random increase of the $X_t$value in relevant steps, given a current value of i. We first concentrate on the contribution of steps of length 1, i.e., we consider $R_1(i) := R(i) \cdot \mathbb{1}\{|R(i)| \leq 1\}$. Thus,

$$\begin{aligned} E(R_1(i)) &= \frac{p_1(i)}{p_{\mathrm{rel}}} - \frac{p_{-1}(i)}{p_{\mathrm{rel}}} \geq \frac{\frac{h}{m^2} \cdot \left(1 - \frac{1}{m}\right)^{m-2}}{\frac{2h+2}{m^2}} - \frac{\frac{2}{m^2} \cdot \left(1 - \frac{1}{m}\right)^{m-2} + \frac{3}{m^4}}{\frac{2h+2}{m^2}} \\ &= \frac{(h-2)\left(1 - \frac{1}{m}\right)^{m-2}}{2h+2} - \frac{3}{m^2(2h+2)} \geq \frac{1}{8 \cdot e} - O(m^{-2}) \end{aligned}$$

since $h \geq 3$. The unconditional decrease $\Delta^-_{>1}(i) = -\Delta(i) \cdot \mathbb{1}\{\Delta(i) < -1\}$, for negative steps of length greater than 1, is in expectation at most

$$E(\Delta^-_{>1}(i)) \leq \sum_{j=2}^{\infty} j \cdot p_{-j}(i) \leq \sum_{j=2}^{\infty} j \cdot (j+1) \cdot \frac{1}{m^{2j}}$$

$$\leq \frac{6}{m^4} + \sum_{j=3}^{\infty} \frac{2m^2}{m^{2j}} = O(m^{-4})$$

using $p_{-j} \leq (j+1)/m^{2j}$. Hence, the total conditional drift is

$$E(R(i)) \geq E(R_1(i)) - \frac{E(\Delta^-_{>1}(i))}{p_{\text{rel}}} \geq \frac{1}{8 \cdot e} - O(m^{-2}) - O(m^{-4}) \cdot em^2$$

$$= \frac{1}{8 \cdot e} - O(m^{-2}),$$

which is bounded from below by a constant such that the first condition of Theorem 4.9 has been established.

The second condition follows with $\delta = 1$ and $r = 8$ from

$$\frac{p_{-j}}{p_{\text{rel}}} \leq \min\left\{1, \frac{j+1}{m^{2j}} \cdot em^2\right\} \leq \min\left\{1, \frac{1}{m^{2j-7}}\right\} \leq 8 \cdot \left(\frac{1}{2}\right)^j$$

for $m \geq 2$. From Theorem 4.9, the lemma follows.   □

We summarize our results. Note that the exponentially small failure probability $2^{-\Omega(\ell)} = 2^{-\Omega(m^{1/3})}$ from Lemma 6.15 is captured by the $O(1/m)$ term of the following lemma.

Theorem 6.17. *Starting with an almost perfect matching and an augmenting path of length $2k+1$, the probability that (1+1) EA_b finds the perfect matching within $2^{c\ell}$ steps, $c > 0$ an appropriate constant, is bounded from above by $O(1/m) + ((2/h) + O(1/m))^k$ if $3 \leq h \leq \ell$ and $k \geq 1$. For $RLS_b^{1,2}$, the bound $2^{-\Omega(\ell)} + (2/h)^k$ holds.*

So far, we have only considered the case of almost perfect matchings and shown that it can take exponential time to achieve the final improvement. We return to the question of whether an almost perfect matching will be reached.

Lemma 6.18. *If (1+1) EA_b or $RLS_b^{1,2}$ do no start with the perfect matching, an almost perfect matching is constructed before the perfect matching with a probability of $\Omega(1/h)$.*

Proof. Let M denote the set of edges selected by the current search point, and let $d := |M \oplus M^*|$ denote the Hamming distance to M^*. We investigate the situations when M is neither an almost perfect nor the perfect matching; this includes the case where M is not even a matching. Then, any step producing an almost perfect matching will be accepted.

For $(1+1)$ EA_b, the probability of producing M^* in the next step is $\Theta(1/m^d)$. We argue that this probability is at most by a factor of $O(h)$ larger than the probability of producing an almost perfect matching in the next step. If $M \oplus M^*$ contains at least one M^* edge, this edge is not included in M. Then the step where everything works as in the step creating the perfect matching, except for the M^* edge, produces an almost perfect matching. The probability $\Theta(1/m^{d-1})$ of this step is even larger than the probability of the step creating M^*. If $M \oplus M^*$ contains no M^* edge, all M^* edges are included in M, and there are $|M^*|$ ways to produce an almost perfect matching by additionally flipping an M^* edge. Their probability is $\Theta(|M^*|/m^{d+1}) = \Theta(1/(hm^d))$. The ratio of the relevant probabilities is always at least $\Omega(1/h)$.

For $\text{RLS}_\text{b}^{1,2}$, a necessary event is a situation where $d \leq 2$. We argue that in any situation where $d = 1$, the next step produces M^* with a probability that is at most by a factor $O(h)$ larger than the probability that it produces an almost perfect matching. In situations where $d = 2$, the first probability will be proven to be even smaller than the last probability since we investigate the next two steps.

Let us consider the case $d = 1$. Then we are only interested in the case where M is a superset of M^* since otherwise M would be almost perfect. Let $M = M^* \cup \{e\}$, implying that e is an $\overline{M}^*$ edge. The next step produces M^* with probability $\Theta(1/m)$. If e and another edge of M flip, an almost perfect matching is obtained. This happens with probabiliy $\Theta(|M^*|/m^2) = \Theta(1/(hm))$. The ratio is $\Omega(1/h)$.

Finally, assume $d = 2$. Then a necessary event to produce M^* is that each of the two edges in $M \oplus M^*$ flips at least once in the next two steps. The probability of this event is $\Theta(1/m^2)$. If $M \oplus M^*$ contains two M^* edges, both are free, and the first step produces an almost perfect matching with a probability of $\Theta(1/m)$ by flipping only one of these edges. If $M \oplus M^*$ contains one M^* edge and one $\overline{M}^*$ edge, the first step removes the latter edge from M with probability $\Theta(1/m)$ and produces an almost perfect matching. Finally, if $M \oplus M^*$ contains two $\overline{M}^*$ edges then $M = M^* \cup \{e_1, e_2\}$ is a non-matching where e_1 and e_2 are the two $\overline{M}^*$ edges. Any step flipping e_1 and an arbitrary M^* edge in the first step will be accepted even though it leads still to a non-matching. The reason is that the penalty term in the underlying fitness function (cf. Section 6.1) decreases by at least 1 and optimization proceeds in lexicographic order. If the second step flips e_2, an almost perfect matching is obtained. Ths probability of these events is $\Theta((|M^*|/m) \cdot (1/m)) = \Theta(1/(hm^2))$, and the ratio of the relevant probabilities is again bounded from below by $\Omega(1/h)$. □

Taking the previous lemma, Lemma 6.12, and Theorem 6.17 together, we obtain that the $2^{\Omega(\ell)}$ bound of the theorem holds with a probability of $\Omega(1/(hm))$ if we start with any search point which is not the optimum. If $h \leq \ell$ then it holds that $\ell = \Omega(m^{1/3})$, and if h is a constant, then $\ell = \Omega(m)$. Altogether, given a initial search point that is not the optimum, we obtain an exponential lower bound of $2^{\Omega(\ell)} = 2^{\Omega(m^{1/3})}$ for the expected optimization time. This is summarized by the following theorem.

Theorem 6.19. *For $G_{h,\ell}$, $3 \leq h \leq \ell$, the expected optimization time of $RLS_b^{1,2}$ and (1+1) EA_b is $2^{\Omega(\ell)}$ if the initial search point is not the perfect matching.*

For example, if the initial search point is drawn uniformly at random, the probability of not starting with the perfect matching is $1 - 2^{-\Omega(m)}$. In general, the precondition of not starting with the optimum is the weakest condition one can think of. Early analyses of simulated annealing for the maximum matching problem (Sasakik and Hajek, 1988) are based on the deterministic choice of the empty matching as initial starting point. Theorem 6.19 is far less restrictive in this sense.

However, Theorem 6.19 contains a result on the expected optimization time, only. This statement goes back to Lemma 6.12 and Theorem 6.17, which imply a lower bound of $\Omega(1/(hm))$ on the probability of observing an exponential optimization time. Giel and Wegener (2004) improve upon this bound and show that an exponential time holds with probability $1 - 2^{-\Omega(\ell)}$ if $h = \omega(\log n)$. Very careful analyses are required to show these improved results, and the interested reader is referred to the works by Giel and Wegener (2003, 2004, 2006).

Conclusions

In this chapter, we have analyzed the simple search algorithms $RLS_b^{1,2}$ and (1+1) EA for the maximum matching problem. Optimal solutions are found on simple graph classes like paths in expected polynomial time. More generally, solutions that are only by a factor $1 + \epsilon$ away from optimality can be found in expected polynomial time. This proves that the algorithms are polynomial-time randomized approximation schemes (PRASs) for the problem. Consistently with this result, the limits of the search algorithms have been determined. On a worst-case graph, the expected time until the optimal solution is found was proven to be exponential.

The analyses make use of the techniques presented in Section 4.2. Most notably, the gambler's ruin theorem and the drift theorem were used to investigate the stochastic processes behind the algorithms.

7

Makespan Scheduling

In this chapter, we study the simple scheduling problem introduced in Section 2.1. Given n jobs with positive processing times $p_1, \ldots, p_n$, schedule them on two identical machines in a way such that the makespan, i.e., the overall completion time, is minimized. Let $x \in \{0,1\}^n$ be a decision vector. Job j is scheduled on machine 1 iff $x_j = 0$ holds and on machine 2 iff $x_j = 1$ holds. Hence, the goal is to minimize

$$f_{p_1,\ldots,p_n}(x) := \max \left\{ \sum_{i=1}^{n} p_j x_j, \sum_{i=1}^{n} p_j (1 - x_j) \right\},$$

where the index $p_1, \ldots, p_n$ is often omitted for the sake of readability. Note that the representation is redundant in the sense that a search point x and its bitwise binary complement $\bar{x}$ lead to the same f value.

We see that the problem is very easy to describe and leads to a pseudo-boolean fitness function in a natural way. In the domain of theoretical computer science, this is a very well-studied problem also known as PARTITION: a set of n numbers has to be split into two subsets such that the numbers in the two subsets sum up to a maximum value as small as possible. In the best case, there is a perfect partition with value $(p_1 + \cdots + p_n)/2$. In the following, we also refer to our scheduling problem briefly by the name PARTITION.

Despite its simplicity, PARTITION is an NP-hard problem with an NP-complete decision variant. Hence, we cannot hope for exact solutions in polynomial time. However, the problem is perfectly suited for investigating the capabilities of stochastic search algorithms to approximate optimal solutions. There are efficient approximation algorithms based on the knapsack problem which guarantee solutions with an approximation ratio $1 + \epsilon$ in time $O(n^3/\epsilon)$, i.e., polynomial in n and $1/\epsilon$ (Hochbaum, 1997). Of course, an approximation ratio of 2 is trivially obtained by placing all jobs on the same machine.

In this chapter, we investigate the simple search algorithms RLS_b^1 and $(1+1)\,\mathrm{EA}_b$ for the minimization of functions $f_{p_1,\ldots,p_n}$ induced by instances to the PARTITION problem. The results are due to Witt (2005). We will start

F. Neumann, C. Witt, *Bioinspired Computation*
in Combinatorial Optimization, Natural Computing Series,
DOI 10.1007/978-3-642-16544-3_7, © Springer-Verlag Berlin Heidelberg 2010

in Section 7.2 with a worst-case perspective, which means that the maximum approximation ratio obtainable in polynomial time over all possible instances is elaborated on. Later, we will relax this perspective by appropriate average-case models, which will be dealt with in Section 7.3.1. Both sections rely on concepts and tools developed in the following section.

7.1 Representations and Neighborhood Structure

Like many of the analyses in previous chapters, we focus on the progress that our stochastic search algorithms achieve by local steps. Our aim is to characterize search points having better Hamming neighbors. To this end, some notions and notations are helpful.

Given the processing times of the n jobs, we assume w. l. o. g. throughout this section that they appear sorted in nonincreasing order, i.e., $p_1 \geq \cdots \geq p_n$. Moreover, we denote by $P := p_1 + \cdots + p_n$ the sum of all processing times, which rephrases our lower bound on optima by $P/2$. As long as a current point $x \in \{0, 1\}^n$ leads to a worse $f$value, the two machines have different loads. We briefly refer to the machine with higher load as the *fuller* machine and call the other machine *emptier*. A local step of a search algorithm might improve a current solution by shifting a job from the fuller to the emptier machine. We describe sufficient conditions for such steps to be successful.

Suppose that a current solution x is given and we know that there exists a job, say job i, on the fuller machine with processing time p_i. If $f(x) \geq P/2 + p_i/2$ holds for this search point, the loads of the machines differ by at least p_i. Hence, a step shifting job i from the fuller to the emptier machine will be accepted. More generally, for an arbitrary search point x, let $s(x)$ denote the smallest processing time of the jobs scheduled on the fuller machine. We call $s(x)$ the *critical job size* with respect to x (and, implicitly, the underlying instance to the PARTITION problem). Our sufficient condition for improvable solutions now reads as $f(x) \geq P/2 + s(x)/2$; see Figure 7.1 for an illustration. Only if the current $f$value is less than this bound can the search algorithm be stuck in a local optimum. Therefore, $P/2 + s(x)/2$ is a possible barrier for locally searching algorithms, and it would be nice to have upper bounds on $s(x)$.

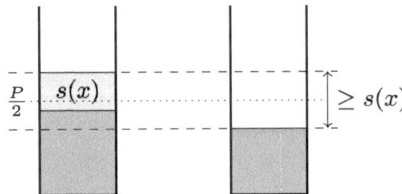

Fig. 7.1. A sufficient condition for locally improvable solutions

In general, given an arbitrary instance to the problem and an arbitrary search point x, there can be no better bound on $s(x)$ than p_1, i.e., the processing time of the largest job. If the processing times of the jobs differ heavily, $P/2 + p_1/2$ might be very close to P; hence our characterization of improvable situations would be trivial. However, after the algorithm has taken some improving steps and has obtained an already relatively good – but not optimal – solution, we might be able to conclude that there must be some quite small jobs on the fuller machine, leading to much better estimates on the critical job size $s(x)$. Actually, we will describe situations where the critical job size is guaranteed to respect a certain bound for the rest of the optimization process. This will be possible since RLS_b^1 and $(1+1)$ EA_b do not accept worsenings.

Given a bound s^* on the critical job size that is maintained for the rest of the optimization, we can estimate the progress of the search algorithms towards the possible barrier $P/2 + s^*/2$ using the techniques from Section 4.2. We allow for barriers $L \geq P/2$ in a slightly more general setting and bound the expected time to reach that barrier using the fitness-level method (Lemma 4.1). Moreover, we present an improved bound for the case where solutions slightly worse than the barrier are sufficient. In that case, the method of expected multiplicative distance decrease (see Section 4.2.3) gives very good time bounds.

Lemma 7.1. *Let a current search point of RLS_b^1 or $(1+1)$ EA_b on an arbitrary instance to the PARTITION problem be given. Suppose that the critical job size is guaranteed to be bounded from above by s^* for all following search points of value greater than $L + s^*/2$. Then the algorithm reaches an f value at most $L + s^*/2$ in expected time $O(n^2)$.*

Proof. The proof uses a fitness-based partition which has to be defined carefully. In particular, since we are dealing with minimization, we now assume $A_i >_f A_{i+1}$ for the sets of the partition.

Let r be the smallest i such that $p_i \leq s^*$, i.e., thanks to $p_1 \geq \cdots \geq p_n$ it holds that $p_i \leq s^*$ for all $i \geq r$. We define

$$A_i := \left\{ x \;\middle|\; P - \sum_{j=r}^{r+i-1} p_j \;\geq\; f(x) \;>\; P - \sum_{j=r}^{r+i} p_j \right\}$$

for $0 \leq i \leq n - r$ and $A_{n-r+1} := \{ x \mid f(x) \leq P - \sum_{j=r}^{n} p_j \}$. Some sets might contain search points of value less than $L + s^*/2$, which we pessimistically ignore in the following. Using the expected time until set A_{n-r+1} is reached, we obtain an upper bound on the expected time to reach the barrier $L + s^*/2$.

Consider some x such that $f(x) > L + s^*/2$. By our assumptions, there must be a job from $p_r, \ldots, p_n$ on the fuller machine whose move to the emptier machine decreases the f value by its processing time or leads to an f value of at most $L + s^*/2$. If $x \in A_i$, there is, thanks to $p_r \geq \cdots \geq p_n$, even a job from $p_r, \ldots, p_{r+i}$ with this property. Moving this job to the emptier machine by a

1-bit flip of the search algorithm has probability at least $1/(en)$ and, again due to $p_r \geq \cdots \geq p_n$, leads to some $x' \in A_j$ such that $j > i$. The expected waiting time for such a step is at most en. After at most $n - r + 1$ sets are left, the $f$value drops to at most $L + s^*/2$. Hence, the total expected time is $O(n^2)$. □

We now turn to a result that is obtained by the method of expected multiplicative distance decrease.

Lemma 7.2. *Let a current search point of RLS_b^1 or (1+1) EA_b on an arbitary instance to the PARTITION problem be given. Suppose that the critical job size is guaranteed to be bounded from above by s^* for all following search points of value greater than $L + s^*/2$. Then for any $\gamma > 1$ and $0 < \delta < 1$, (1+1) EA_b (RLS_b^1) reaches an $f$value at most $L + s^*/2 + \delta P/2$ in at most $\lceil en \ln(\gamma/\delta)\rceil$ ($\lceil n \ln(\gamma/\delta)\rceil$) steps with probability at least $1 - \gamma^{-1}$. Moreover, the expected number of steps is at most $2\lceil en \ln(2/\delta)\rceil$ ($2\lceil n \ln(2/\delta)\rceil$).*

Proof. Let r be the smallest i such that $p_i \leq s^*$. Moreover, consider a current search point x satisfying $f(x) > L + s^*/2$. We are interested in the contribution of the so-called small jobs $p_r, \ldots, p_n$ to the $f$value and estimate the average decrease of the $f$value using the method of expected multiplicative distance decrease.

Let $d(x) := \max\{f(x) - L - s^*/2, 0\}$ denote the distance of the $f$value from the barrier $L + s^*/2$. By our assumptions, $d(x)$ is a lower bound on the contribution of small jobs to $f(x)$. Moreover, $f(x) \geq P/2$ and $f(x) > L + s^*/2$ together imply $d(x) \leq P/2$. As long as $d(x) > 0$, all steps moving only a small job to the emptier machine are accepted and decrease the $d$value by its size or lead to an $f$value of at most $L + s^*/2$. Let d_0 be some current $d$value. Since a 1-bit flip of (1+1) EA_b has probability at least $1/(en)$, the expected $d$decrease is at least $d_0/(en)$ and the expected $d$value after the step, therefore, is at most $(1 - 1/(en))d_0$. The expected $d$value d_t after t steps is at most $(1 - 1/(en))^t d_0$. For $t' := en \ln(\gamma/\delta)$, we have $d_{t'} \leq \delta d_0/\gamma \leq \delta P/(2\gamma)$. Markov's inequality implies that $d_{t'} \leq \delta P/2$ with probability at least $1 - 1/\gamma$. Moreover, we can repeat independent phases of length $\lceil en \ln(2/\delta)\rceil$. The expected number of phases until the $d$value is at most $\delta P/2$ is at most 2, implying the lemma for (1+1) EA_b.

The statements regarding RLS_b^1 follow in the same way, taking into account that a 1-bit flip has probability $1/n$. □

7.2 Worst-Case Analysis

7.2.1 Approximations Obtainable in Expected Polynomial Time

In this section, we will study bounds on the approximation ratios obtainable by the search algorithms within polynomial time regardless of the problem instance. This is a classical worst-case analysis, and we cannot hope for exact solutions in polynomial time due to the NP-hardness of the problem at hand.

Theorem 7.3. *Let $\epsilon > 0$ be a constant. On every instance to the PARTITION problem, (1+1) EA_b and RLS_b^1 reach an f value with approximation ratio $4/3+$ ϵ in an expected number of $O(n)$ steps and an f value with approximation ratio $4/3$ in an expected number of $O(n^2)$ steps.*

Proof. We define $L := \max\{p_1, P/2\}$ and are interested in an upper bound on the smallest job on the fuller machine. To this end, we still have to distinguish between two cases. The first case holds if $p_1 + p_2 > 2P/3$. Recalling $p_1 \geq \cdots \geq p_n$, this implies $p_1 > P/3$ and, therefore, $P - p_1 < 2P/3$. Hence, if we start with p_1 and p_2 on the same machine, a step separating p_1 and p_2 by putting p_2 onto the emptier machine is accepted, and these jobs will remain separated afterwards. The expected time until such a separating step occurs is $O(n)$. We claim that for all following search points x of value at least $f(x) > L$, the critical job size is bounded from above by p_3. This holds since by the definition of L, the biggest job is not sufficient for obtaining an f value greater than L. Now, since $p_3 + \cdots + p_n < P/3$, we know that $p_i < P/3$ for $i \geq 3$. When working with $s^* \leq P/3$, search points of value $L + s^*/2$ have an approximation ratio of

$$\frac{L + s^*/2}{L} \leq \frac{P/2 + s^*/2}{P/2} \leq 1 + \frac{P/6}{P/2} = \frac{4}{3}$$

since $L \geq P/2$. Likewise, search points of value $L + s^*/2 + \delta P/2$ have an approximation ratio of $4/3 + \delta$. Hence, the first statement of the theorem follows for $\delta := \epsilon$ by Lemma 7.2 and the second one by Lemma 7.1.

If $p_1 + p_2 \leq 2P/3$, we have $p_i \leq P/3$ for $i \geq 2$. Since $p_1 < P/2$, this implies that the critical job size is always at most $p_2 \leq P/3$. Therefore, the theorem holds also in this case. □

The approximation ratio $4/3$ that the search algorithms are able to obtain within expected polynomial time is at least almost tight. We present a simple worst-case instance where both RLS and (1+1) EA_b get stuck at approximation ratios close to $4/3$ with probability $\Omega(1)$. This instance is called P_ϵ^* in the following.

Definition 7.4. *Let n be even and $\epsilon > 0$ be an arbitrarily small constant. Then the instance $P_\epsilon^* = \{p_1, \ldots, p_n\}$ is defined by $p_1 := p_2 := 1/3 - \epsilon/4$ and $p_i := (1/3 + \epsilon/2)/(n - 2)$ for $3 \leq i \leq n$.*

Note that the total processing time $P := p_1 + \cdots + p_n$ has been normalized to 1 for the instance P_ϵ^*, which is only a cosmectic aspect and does not play a crucial role. However, it is important that the processing times in the instance be highly diverse: the times $p_1 = p_2$ of the two big jobs are almost as large as the total processing time of all $n - 2$ small jobs.

It is worth noting that the instance P_ϵ^* has an exponential number of perfect partitions. Each solution that puts one big job and half of the small jobs on each machine is such a perfect partition. However, the difference in job sizes can trick the search algorithm into bad local optima as follows: suppose the two big jobs are on one machine and all the small jobs on the other one. Then the current makespan is $p_1 + p_2 = 2/3 - \epsilon/2$. A step that tries to move a big job from the fuller to the emptier machine would make the previously emptier machine have a makespan of $(1/3 + \epsilon/2) + 1/3 - \epsilon/4 = 2/3 + \epsilon/4$. This is worse than the previous makespan by an amount of $3\epsilon/4$. To compensate for this, one would need at least $(3\epsilon/4)/p_3 = \Omega(n)$ small jobs. An illustration is given in Figure 7.2.

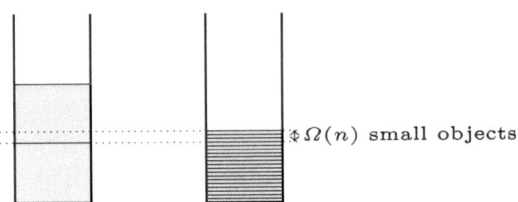

$\Omega(n)$ small objects

Fig. 7.2. A worst-case instance

In the following theorem, we make our ideas precise. As a technical detail, we do not wait for all small jobs to be scheduled on the emptier machine but for only for almost all of these.

Theorem 7.5. *Let ϵ be any constant s. t. $0 < \epsilon < 1/3$. With probability $\Omega(1)$, both (1+1) EA_b and RLS_b^1 need on the instance P_ϵ^* at least $n^{\Omega(n)}$ steps to create a solution with a better approximation ratio than $4/3 - \epsilon$.*

Proof. The proof idea is to show that the search algorithm reaches a situation where the two big jobs are one the same machine and at least $k := n - 2 - (n-2)\epsilon/2$ small jobs are on the other one. Since $\epsilon < 1/3$, at least k jobs yield a total processing time of at least $1/3 + \epsilon/2 - (\epsilon/2)(1/3 + \epsilon/2) = 1/3 + \epsilon/3 - \epsilon^2/4 \geq 1/3 + \epsilon/4$. To leave the situation by separating the big jobs, the search algorithm has to transfer small jobs of a total processing time of at least $\epsilon/4$ from the emptier to the fuller machine in a single step. For this, $(n-2)\epsilon/2$ small jobs are not enough. Flipping $\Omega(n)$ bits in one step of (1+1) EA_b has probability $n^{-\Omega(n)}$, and flipping $\Omega(n)$ bits at least once within n^{cn} steps is, therefore, still exponentially unlikely if the constant c is small enough. For RLS_b^1, the probability is even 0. The makespan is at least $2/3 - \epsilon/2$ unless the two big jobs are separated, which corresponds to an approximation ratio no better than $(2/3 - \epsilon/2)/(1/2) = 4/3 - \epsilon$. This will imply the theorem if we can prove that the described situation is reached with probability $\Omega(1)$.

The open claim is shown by considering the initial search point of the search algorithm. With probability $1/2$, it puts the two big jobs onto the same machine. Therefore, we estimate the probability that enough small jobs are transferred from this machine to the other one in order to reach the situation, before a bit at the first two positions (denoting the large jobs) flips. In a phase of length cn for any constant c, with probability $(1 - 2/n)^{cn} = \Omega(1)$, the latter never happens. Under this assumption, each step moving a small job onto the emptier machine is accepted. By the same idea as that in the proof of Lemma 7.2, we estimate the expected decrease of the contribution of small jobs to the $f$value. Reducing it to at most an $\epsilon/2$fraction of its initial contribution suffices for obtaining at least k jobs in the emptier machine. Each step leads to an expected decrease by at least a $1/(en)$fraction. Since ϵ is a positive constant, $O(n)$ steps are sufficient to decrease the contribution to at most an expected $\epsilon/4$fraction. By Markov's inequality, we obtain the desired fraction within $O(n)$ steps with probability at least $1/2$. Since c may be chosen appropriately, this proves the theorem. □

7.2.2 The Success Probability for Certain Approximations

The worst-case example P_ϵ^* studied in the previous subsection suggests that the search algorithms are likely to arrive at a bad approximation if they misplace big jobs. If we take the worst-case perspective on what is doable in expected polynomial time then the probability $\Omega(1)$ of getting stuck in a local optimum as proved in Theorem 7.5 necessarily limits our result to the relatively bad approximation ratio of only $4/3 - \epsilon$.

However, with similar techniques as before, it can easily be shown for P_ϵ^* that the search algorithms are able to find an optimal solution with probability $\Omega(1)$ in polynomial time if they separate the two big jobs in the beginning. This is a finer result from a more relaxed perspective and relates to the success probability within polynomial time. Even if the expected optimization time of a search algorithm is exponential, it might have a good probability of finding optima in polynomial time (Droste et al., 2002). Multiple restarts of the search algorithms will help us find the optimum within polynomial time with a probability very close to 1.

An obvious question is whether the observation regarding the success probability on P_ϵ^* can be generalized to arbitrary instances to the PARTITION problem. We can achieve this in the following way. In order to obtain a $(1 + \epsilon)$ approximation in polynomial time according to Lemma 7.1, the critical job size should be bounded above by ϵP. Due to the ordering $p_1 \geq \cdots \geq p_n$, all objects of index at least $s := \lceil 1/\epsilon \rceil$ are bounded by this volume. Let those jobs be called small and the first $s - 1$ jobs be called large; see Figure 7.3 for an illustration. The idea is to bound the probability that the search algorithm distributes the large jobs in such a nice way that it makes mistakes only with the small jobs, resulting in a critical job size of at most p_s. Interestingly, this is essentially the same idea as that for the classical approximation (PTAS, see Definition 2.7) scheme for the PARTITION problem presented by Graham (1969).

Hence, as long as ϵ does not depend on the input size, we can achieve almost arbitrarily good approximations in polynomial time. In the following,

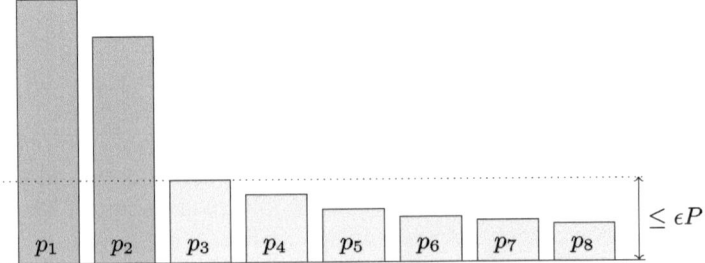

Fig. 7.3. Example of large and small jobs, $s = 3$

we will show that the search algorithms are able to achieve similar results with certain properties. Actually, the analysis is based on the above-described distinction of large and small jobs and a simulation of Graham's PTAS. Even if the search algorithm does not know the latter's algorithmic idea, it is able to behave accordingly by chance.

Theorem 7.6. *Let an arbitrary instance to the PARTITION problem be given and choose* $\epsilon \geq 4/n$. *Then, with probability at least* $2^{-(e\log e + e)\lceil 2/\epsilon\rceil \ln(4/\epsilon) - \lceil 2/\epsilon\rceil}$, *(1+1) EA$_b$ creates a solution of approximation ratio* $(1 + \epsilon)$ *in* $\lceil en\ln(4/\epsilon))\rceil$ *steps. The same holds for* RLS_b^1 *with* $\lceil n\ln(4/\epsilon))\rceil$ *steps and a probability of at least* $2^{-(\log e + 1)\lceil 2/\epsilon\rceil \ln(4/\epsilon) - \lceil 2/\epsilon\rceil}$.

Proof. Let $s := \lceil 2/\epsilon \rceil \leq n/2 + 1$. Since $p_1 \geq \cdots \geq p_n$, it holds that $p_i \leq \epsilon P/2$ for $i \geq s$. If $p_1 + \cdots + p_{s-1} \leq P/2$, the critical job size for all search points of value at least $P/2 + p_s/2$ is always bounded above by p_s and, therefore, by $\epsilon P/2$. Therefore, in this case, the theorem follows for $\delta := \epsilon/2$ and $\gamma := 2$ by Lemma 7.2.

In the following, we assume $p_1 + \cdots + p_{s-1} > P/2$. Consider all partitions of only the first $s - 1$ jobs, i.e., the large jobs. Let L^* be the minimum makespan over all these partitions and let $L := \max\{P/2, L^*\}$. A search point and its complement lead to the same f-value. Hence, there at least two search points such that the contribution of the large jobs to the makespan is at most L. Since the initial solution is drawn uniformly, the probability is at least 2^{-s+2} that the big jobs in the inital solution contribute at most L to the makespan.

As long as the big jobs are not moved, we can be sure that the critical job size for search points of f-value greater than L is at most $p_s \leq \epsilon P/2$, and we can apply the arguments from the first paragraph. The probability that in a phase of $t := \lceil en\ln(4/\epsilon)\rceil$ steps it never happens that at least one of the first $s - 1$ bits flips is bounded from below by

$$\left(1 - \frac{s-1}{n}\right)^{en(\ln(4/\epsilon))+1} \geq e^{-e(\ln(4/\epsilon))(s-1)}\left(1 - \frac{s-1}{n}\right)^{se\ln(4/\epsilon)},$$

which is at least $2^{-(e\log e + e)\lceil 2/\epsilon\rceil \ln(4/\epsilon)}$ since $s - 1 \leq n/2$. Assuming this to happen, we apply Lemma 7.2 for $\delta := \epsilon/2$ and $\gamma := 2$. Hence, (1+1) EA$_b$

reaches a solution of approximation ratio $(1 + \epsilon)$ within t steps with probability at least $1/2$. Altogether, the desired approximation is reached within t steps with probability at least

$$\frac{1}{2} \cdot 2^{-\lceil 2/\epsilon \rceil + 2} \cdot 2^{-(e \log e + e)\lceil 2/\epsilon \rceil \ln(4/\epsilon)} \geq 2^{-(e \log e + e)\lceil 2/\epsilon \rceil \ln(4/\epsilon) - \lceil 2/\epsilon \rceil}.$$

The statement for RLS_b^1 follows by redefining $t := \lceil n \ln(4/\epsilon) \rceil$. □

Theorem 7.6 allows us to design a PRAS (see Definition 2.8) for the PARTITION problem using multistart variants of the considered search algorithms. The idea is as follows. If $\ell(n)$ is a lower bound on the probability that a single run of the algorithm achieves the desired approximation in $O(n \ln(1/\epsilon))$ steps, then this holds for at least one out of $\lceil 2/\ell(n) \rceil$ parallel runs with a probability of at least

$$1 - (1 - \ell(n))^{\lceil 2/\ell(n) \rceil} \geq 1 - e^{-2} > 3/4,$$

which is the success probability required in a PRAS.

We terminate each run definitely after $O(n \ln(1/\epsilon))$ steps. Hence, the computational effort $c(n)$, i.e., the number of $f$evaluations, incurred by the parallel runs can be bounded according to

$$O(n \ln(1/\epsilon)) \cdot 2^{(e \log e + e)\lceil 2/\epsilon \rceil \ln(4/\epsilon) + O(1/\epsilon)}.$$

For $\epsilon > 0$ a constant, $c(n) = O(n)$ holds, which is a polynomial in n. Moreover, $c(n)$ is still a polynomial for any $\epsilon = \Omega(\log \log n / \log n)$. Hence, the multistart strategy is really a PRAS. This is the second example where a stochastic search algorithm is characterized in such a way. Compared to the result for the maximum matching problem in Section 6.2, the statement from this section might be considered even more promising since we are dealing with an NP-hard problem.

7.3 Average-Case Analysis

7.3.1 Introductory Results

So far, the analyses in this chapter were based on a pessimistic model. All possible instances to the problem had to be taken into account, resulting in all statements dealing implicitly with the worst case. It is commonly objected that this perspective of worst-case instances might be very unlikely in applications or that they do not even appear at all.

A well-established relaxation to the worst-case perspective is a so-called average-case analysis. The average is taken of a set of instances, each of which obtains a certain probability of occurrence, in the simplest case uniform over the considered set. In such a model, even the behavior of a deterministic algorithm has to be analyzed in a stochastic environment. A classical example is the average-case analysis of QuickSort (Cormen et al., 2001), where each initial order of the objects to be sorted is equally likely and an expectation of the runtime is computed. The worst-case runtime $\Omega(n^2)$ in the deterministic case decreases to an expectation of only $O(n \log n)$ in the average-case model.

We aim at an average case of stochastic search algorithms, which entails two sources of randomness. We have to deal with random inputs and random decisions of the search algorithm at the same time. This might explain why average-case analyses of stochastic search algorithms are relatively rare and our attempt is one of the first such analyses. As a result, we must restrict ourselves to fairly simple distributions on the set of possible instances.

Two distributions are considered:

Uniform distribution model, where each job size p_i is drawn independently from the unit interval $[0, 1]$.

Exponential distribution model, where each job size p_i independently follows an exponential distribution with parameter 1, i.e., $\text{Prob}(p_i \geq t) = e^{-t}$ for $1 \leq i \leq n$.

Using the properties of the two distributions, the expected job size is $1/2$ in the first and 1 in the second model. It is crucial for our analyses that we assume independence of the random job sizes. We also drop the assumption that the jobs appear sorted in decreasing size since it would introduce dependencies. If we change our perspective from random job sizes p_i to a sorted representation, this will be made explicit in the following.

Our two models slightly abuse the definition of the PARTITION problem in that the random job sizes are positve reals rather than integers now. By limiting the precision of the numbers and normalizing the job sizes, however, we can easily arrive at an instance according to the original formulation. This is a technical detail that will not be discussed hereinafter.

A different motivation for our average-case models is taken from the literature. In the last two decades, some average-case analyses of *deterministic* algorithms for the PARTITION problem have been performed. The first such analyses studied the LPT *(longest processing time)* rule, a greedy algorithm displayed as Algorithm 10.

Algorithm 10 Longest Processing Time (LPT)

---

1. Sort the jobs according to decreasing processing time.
2. For $i = 1, \ldots, n$, schedule the ith job on the currently emptier machine (breaking ties arbitrarily)

Extending a result that showed convergence in expectation, Frenk and Rinnooy Kan (1986) were able to prove that the LPT rule converges to optimality at a speed of $O(\log n/n)$ almost surely in several input models, including the above-mentioned uniform distribution and exponential distribution models. Further results on average-case analyses of more elaborate deterministic algorithms are contained in a survey by Coffman and Whitt (1995).

In our random input models, the optimum f value is itself random and an appropriate generalization of the approximation ratio is no longer obvious. Therefore, for a current search point, we now consider the so-called *discrepancy* measure also studied by Frenk and Rinnooy Kan (1986). The discrepancy denotes the absolute difference of the loads of the two machines; formally, we have $|f(x) - f(\bar{x})|$ as the discrepancy of a search point x. As a first observation, we show that the random initial search point is likely to have a relatively high discrepancy.

Lemma 7.7. *With probabilty $\Omega(1)$, the initial discrepancy of (1+1) EA_b and RLS is $\Omega(\sqrt{n})$ in both the uniform and the exponential distribution models.*

Proof. We show that the discrepancy is a consequence of a bias of the binomial distribution. Let F be the random random number of jobs which are scheduled on the first machine by the initial search point. Since each bit of the initial search point is drawn uniformly and independently, F follows a binomial distribution with parameters n and $1/2$. It is well known (Jansen and Wegener, 2001) that such a random variable is likely to exceed its expectation in the order of its standard deviation; more precisely, the probability of the event $F \geq n/2 + c\sqrt{n}$ is $\Omega(1)$ for a constant $c > 0$. Let us assume this event to happen. Then there are $\Omega(\sqrt{n})$ more jobs on the first than on the second machine, which does not say anything about the discrepancy so far.

Let $p_1^f, \ldots, p_k^f$, where $k = n/2 + c\sqrt{n}$, denote the processing times of the jobs on the first machine (without assuming any specific order), and accordingly $p_1^s, \ldots, p_{n-k}^s$ on the second machine. Note that the initial search point decides on which machine to schedule a job independently of the job size. Hence, we can apply the principle of deferred decisions (Motwani and Raghavan, 1995). This principle means that the outcomes of random variables are only revealed at the time when they are first needed. Here, we assume the processing times to be chosen independently after the initial schedule has been fixed. If we study only $n - k$ of the jobs on the first machine, we can compare their processing times to those of the jobs $p_1^s, \ldots, p_{n-k}^s$. Since no specific order is assumed, $p_1^f + \cdots + p_{n-k}^f$ follows the same distribution as $p_1^s + \cdots + p_{n-k}^s$. By symmetry, $p_1^f + \cdots + p_{n-k}^f \geq p_1^s + \cdots + p_{n-k}^s$ holds with probability $\Omega(1)$. If we can show that $p_{n-k+1}^f + \cdots + p_k^f = \Omega(\sqrt{n})$, we have proved the lemma.

The last claim follows in almost the same way for both random models. For $1 \leq i \leq n$, we have $\mathrm{Prob}(p_i \geq 1/2) = 1/2$ in the uniform distribution model and $\mathrm{Prob}(p_i \geq 1/2) \geq e^{-1/2} \geq 1/2$ in the exponential distribution model. Note that there are $\Omega(\sqrt{n})$ jobs in the set $p_{n-k+1}^f, \ldots, p_k^f$. Counting the jobs of size at least $1/2$ among these and applying Chernoff bounds, we obtain that $p_{n-k+1}^f + \cdots + p_k^f = \Omega(\sqrt{n})$ with probability $\Omega(1)$. All assumptions together hold with probability $\Omega(1)$, which proves the lemma. $\square$

We continue by showing a simple upper bound on the discrepancy after polynomially many steps in the uniform distribution model.

Lemma 7.8. *The discrepancy of (1+1) EA_b (RLS_b^1) in the uniform distribution model is bounded from above by 1 after an expected number of $O(n^2)$ ($O(n \log n)$) steps. Moreover, for any constant $c \geq 1$, it is bounded from above by 1 with probability at least $1 - O(1/n^c)$ after $O(n^2 \log n)$ ($O(n \log n)$) steps.*

Proof. Recall the argumentation behind the critical job size defined in Section 7.1. If the discrepancy is greater than 1, steps flipping one bit can improve the f value by the job moved or lead to a discrepancy of less than 1. By a fitness-level argument as in the proof of Lemma 7.1, we obtain the $O(n^2)$ bound for (1+1) EA_b. This holds for any random instance. Hence, by Markov's inequality and repeating phases, the discrepancy is at most 1 with probability $1 - O(1/n^c)$ after $O(n^2 \log n)$ steps. The statements for RLS_b^1 follow immediately by the Coupon Collector's Theorem (see Section 4.2.2).   □

The preceding upper bound on the discrepancy was easy to obtain; however, for (1+1) EA_b, we can show that with a high probability, the discrepancy provably becomes much lower than 1 in a polynomial number of steps. The reason is as follows. All proofs so far considered only local steps; however, (1+1) EA_b is able to leave local optima by flipping several bits in a step. In particular, it can swap two jobs that are on different machines, which changes the makespan by the difference of the two job sizes. This allows for further improvements of the f value until there are no more possible operations swapping two jobs.

Using this observation, we extend the set of jobs used to determine the critical job size by "difference jobs" of size $p_i - p_j$, $1 \leq i, j \leq n$, such that $p_i > p_j$, and p_i is on the fuller and p_j on the emptier machine. The aim is therefore to bound the difference jobs in our random instance. To this end, we finally sort the n random job sizes $p_1, \ldots, p_n$ decreasingly. Let $X_{(1)} \geq \cdots \geq X_{(n)}$ be the resulting sequence, which is now a sequence of dependent random variables. The $X_{(i)}$, $1 \leq i \leq n$, are typically called *order statistics* (David and Nagaraja, 2003). We state useful properties that hold for the order statistic in our models.

Uniform distribution model. For $1 \leq i \leq n - 1$ and $0 < t < 1$ it holds that $\text{Prob}(X_{(i)} - X_{(i+1)} \geq t) = \text{Prob}(X_{(n)} \geq t) = (1 - t)^n$.

Exponential distribution model. For $1 \leq i \leq n$ it holds that $X_{(i)} = \sum_{j=i}^{n} \frac{Y_j}{j}$, where $Y_1, \ldots, Y_n$ is a sequence of independent, exponentially distributed random variables with parameter 1 (the same sequence is used in all $X_{(i)}$).

With respect to the uniform distribution model, the properties immediately allow us to bound the size of the "difference jobs" $X_{(i)} - X_{(i+1)}$ if we know that $X_{(i)}$ is on the fuller and $X_{(i+1)}$ on the emptier machine. With regard to the exponential distribution model, a little bit more work is required to estimate the size of the difference jobs. However, the basic ideas of the upcoming analyses are contained in the proofs for the uniform distribution model.

7.3.2 Asymptotically Vanishing Discrepancies

The following theorem bounds the discrepancy of $(1+1)$ EA_b in the uniform distribution model.

Theorem 7.9. *Let $c \geq 1$ be an arbitrary constant. After $O(n^{c+4} \log n)$ steps, the discrepancy of $(1+1)$ EA_b in the uniform distribution model is bounded from above by $O(\log n / n)$ with probability at least $1 - O(1/n^c)$. Moreover, the expected discrepancy after $O(n^5 \log n)$ steps is also bounded by $O(\log n / n)$.*

Before giving the proof, we try to interpret the result. First, the solution of $(1+1)$ EA_b after a polynomial number of steps converges to optimality *in expectation*. Second, the asymptotic discrepancy after a polynomial number of steps is at most $O(\log n / n)$, i.e., convergent to 0, with probability $1 - O(1/n^c)$, i.e., convergent to 1 polynomially fast. This is almost as strong as the above-mentioned result for the LPT rule proved by Frenk and Rinnooy Kan (1987).

Proof. By Lemma 7.8, the discrepancy is at most 1 after $O(n^2 \log n)$ steps with probability at least $1 - O(1/n^2)$. Since the discrepancy is always bounded by n, the failure probability contributes only an $O(1/n)$ term to the expected discrepancy after $O(n^5 \log n)$ steps. From now on, we consider the time after the first step where the discrepancy is at most 1 and concentrate on steps flipping two bits. If an accepted step moves an object of size p' from the fuller to the emptier machine and one of size $p'' < p'$ the other way round, the discrepancy may be decreased by $2(p' - p'')$. We look for combinations where $p' - p''$ is small.

Let $X_{(1)} \geq \cdots \geq X_{(n)}$ be the order statistics of the random job sizes. If for the current search point there is some i s.t. $X_{(i)}$ is the order statistic of a job on the fuller and $X_{(i+1)}$ on the emptier machine, then a step exchanging $X_{(i)}$ and $X_{(i+1)}$ may decrease the discrepancy by $2(X_{(i)} - X_{(i+1)})$. If no such i exists, all jobs on the emptier machine are larger than every job on the fuller machine. In this case, $X_{(n)}$ can be moved onto the emptier machine, possibly decreasing the discrepancy by $2X_{(n)}$. Hence, we need upper bounds on both $X_{(i)} - X_{(i+1)}$ and $X_{(n)}$.

Let $t^* := (c+1)(\ln n)/n$, i.e., $t^* = O(\log n / n)$ since c is a constant. We obtain $(1 - t^*)^n \leq n^{-c-1}$. By the above-mentioned properties of order statistics, this implies that with probability $1 - O(1/n^c)$, $X_{(i)} - X_{(i+1)} \leq t^*$ holds for all i and $\mathrm{Prob}(X_{(n)} \geq t^*) = O(1/n^{c+1})$. Now assume $X_{(i)} - X_{(i+1)} \leq t^*$ for all i and $X_{(n)} \leq t^*$. If this does not hold, we bound the expected discrepancy after $O(n^{c+4} \log n)$ steps by 1, yielding a term of $O(1/n^c) = O(1/n)$ in the total expected discrepancy. By the arguments explaining the critical job size, there is always a step flipping at most two bits that decreases the discrepancy as long as the discrepancy is greater than t^*.

It remains to estimate the time to decrease the discrepancy. Therefore, we need lower bounds on $X_{(i)} - X_{(i+1)}$ and X_n. Let $\ell^* := 1/n^{c+2}$. We obtain $\mathrm{Prob}(X_{(i)} - X_{(i+1)} \geq \ell^*) \geq \mathrm{e}^{-2/n^{c+1}} \geq 1 - 2/n^{c+1}$. Hence, with probability $1 - O(1/n^c)$, $X_{(i)} - X_{(i+1)} \geq \ell^*$ for all i. Moreover, $X_{(n)} \geq \ell^*$ with probability $1 - O(1/n^{c+1})$. We assume these lower bounds to hold, introducing a failure probability of only $O(1/n^c)$. The contribution of this failure probability to the expected discrepancy is negligible, as above. A step flipping one or two specific bits has probability at least $n^{-2}(1 - 1/n)^{n-2} \geq 1/(\mathrm{e}n^2)$. Hence, the discrepancy is decreased by at least ℓ^* or drops below t^* with probability $\Omega(1/n^2)$ in each step. The expected time until the discrepancy becomes at most t^* is, therefore, bounded from above by $O(\ell^* n^2) = O(n^{c+4})$, and, by repeating phases, the time is at most $O(n^{c+4} \log n)$ with probability $1 - O(1/n^c)$. The sum of all failure probabilities is $O(1/n^c)$. □

We finally elaborate on a result for the exponential distribution model similar in flavor to Theorem 7.9. The line of proof will also be similar, but more arguments are needed to bound the size of the "difference jobs."

Theorem 7.10. *Let $c \geq 1$ be an arbitrary constant. With probability $1 - O(1/n^c)$, the discrepancy of (1+1) EA_b in the exponential distribution model is bounded above by $O(\log n)$ after $O(n^2 \log n)$ steps and by $O(\log n/n)$ after $O(n^{c+4} \log^2 n)$ steps. Moreover, the expected discrepancy is $O(\log n)$ after $O(n^2 \log n)$ steps and it is $O(\log n/n)$ after $O(n^6 \log^2 n)$ steps.*

The proof of the theorem relies on the following probabilistic argument.

Lemma 7.11. *Let S_k denote the sum of k independent, exponentially distributed random variables with parameter 1. Then $\mathrm{Prob}(S_k > 2k) = 2^{-\Omega(k)}$.*

Proof. It is well known (Feller, 1971) that S_k follows a gamma distribution, i.e.,

$$\mathrm{Prob}(S_k \geq 2k) = \mathrm{e}^{-2k}\left(1 + \frac{2k}{1!} + \cdots + \frac{(2k)^{k-1}}{(k-1)!}\right) \leq \frac{k\mathrm{e}^{-2k}(2k)^{k-1}}{(k-1)!}.$$

By Stirling's formula, the last expression is bounded above by

$$\frac{\mathrm{e}^{-2k+(k-1)} \cdot 2^{k-1} \cdot k \cdot k^{k-1}}{(k-1)^{k-1}} = \mathrm{e}^{-2k+(k-1)} \cdot 2^{k-1} \cdot k \cdot \left(1 - \frac{1}{k}\right)^{-(k-1)} = 2^{-\Omega(k)}.\ \square$$

Proof (Theorem 7.10). Each job has an expected size of 1 in the exponential distribution model. At first, this bounds the initial discrepancy trivially by n. In the following, all failure probabilities will be bounded by $O(1/n^2)$. In the case of a failure, we will tacitly bound the contribution of the failure to the expected discrepancy after $O(n^2 \log n)$ or $O(n^6 \log^2 n)$ steps by the expected initial discrepancy multiplied by the failure probability, which yields a contribtuion of $O(1/n)$. Next, we will show that with probability $1 - O(1/n^c)$, the critical job size of all search points is always $O(\log n)$. Together with Lemma 7.1, this claim implies the theorem for the situation after $O(n^2 \log n)$ steps.

To show the claim, we again consider the order statistics $X_{(1)} \geq \cdots \geq X_{(n)}$ of the random job sizes. Our goal is to show that with high probability, $X_{(1)} + \cdots + X_{(k)} \leq P/2$ holds for $k := \lceil \delta n \rceil$ and some constant $\delta > 0$. Afterwards, we will prove that $X_{(k)} = O(\log n)$ with high probability.

Each job in the exponential distribution model has a size of at least 1 with probability $e^{-1} > 1/3$. By Chernoff bounds, $P \geq n/3$ with probability $1 - 2^{-\Omega(n)}$. To bound $X_{(1)} + \cdots + X_{(k)}$, we use the above-mentioned identity $X_{(i)} = \sum_{j=i}^{n} Y_j/j$. Hence,

$$X_{(1)} + \cdots + X_{(k)} = Y_1 + 2 \cdot \frac{Y_2}{2} + \cdots + k \cdot \frac{Y_k}{k} + k \sum_{i=k+1}^{n} \frac{Y_i}{i}$$

$$\leq \sum_{j=1}^{k} Y_j + \sum_{i=1}^{\lceil n/k \rceil} \frac{1}{i} \sum_{j=ik+1}^{(i+1)k} Y_j,$$

where $Y_j := 0$ for $j > n$. Essentially, we are confronted with $\lceil n/k \rceil$ sums of k exponentially distributed random variables each. By Lemma 7.11, a single sum is bounded above by $2k$ with probability $1 - 2^{-\Omega(k)}$, which is at least $1 - 2^{-\Omega(n)}$ for the values of k considered. Since we consider at most n sums, this statement also holds for all sums together. Hence, with probability $1 - 2^{-\Omega(n)}$, the considered expression is bounded above by

$$2\lceil \delta n \rceil + \sum_{i=1}^{1/\delta} \frac{2\lceil \delta n \rceil}{i} \leq 2(\delta n + 1)\ln(1/\delta + 2),$$

which is strictly less than $n/6$ for $\delta \leq 1/50$ and n large enough. Together with the above lower bound on w, this implies that with probability $1 - 2^{-\Omega(n)}$, the critical job size is always bounded above by the $\lceil n/50 \rceil$th largest size.

How large is $X_{(\lceil n/50 \rceil)}$? Since with probability at least $1 - ne^{-(c+1)\ln n} \geq 1 - n^{-c}$, all random variables Y_j are bounded above by $(c+1)\ln n$, it follows that with at least the same probability, we have

$$X_{\lceil n/50 \rceil} = \sum_{j=\lceil n/50 \rceil}^{n} \frac{Y_j}{j} < (c+1)(\ln n)((\ln n) + 1 - \ln(n/49))$$

(for n large enough), which equals $(c+1)(\ln(49) + 1)(\ln n) = O(\log n)$. The sum of all failure probabilities is $O(1/n^c)$, bounding the critical size as desired.

We still have to show the theorem for the case of $O(n^{c+4} \log^2 n)$ steps. Now we assume that the discrepancy has been decreased to $O(\log n)$ and use the same idea as that in the proof of Theorem 7.9 by investigating steps swapping $X_{(i)}$ and $X_{(i+1)}$ or moving $X_{(n)}$. Above, we have shown that with probability $1 - O(1/n^c)$, the smallest job on the fuller machine is always at most $X_{(k)}$ for some $k \geq n/50$. Since $X_{(k)} - X_{(k+1)} = Y_k/k$, we obtain $X_{(k)} - X_{(k+1)} \leq 50 Y_k/n$ with the mentioned probability. Moreover, it was shown that $Y_j \leq (c+1) \ln n$ for all j with at least the same probability. Altogether, $X_{(k)} - X_{(k+1)} \leq 50(c+1)(\ln n/n) =: t^*$ with probability $1 - O(1/n^c)$. Since $X_{(n)} = Y_n/n$, $\text{Prob}(X_{(n)} \leq t^*)$ with probability $1 - O(1/n^c)$, too. In the following, we assume these upper bounds to hold. This implies that as long as the discrepancy is greater than t^*, there is a step flipping at most two bits and decreasing the discrepancy.

It remains to establish lower bounds on $X_{(k)} - X_{(k+1)}$ and $X_{(n)}$. We know that $X_{(k)} - X_{(k+1)} \geq Y_k/n$ and obtain $\text{Prob}(X_{(k)} - X_{(k+1)} \geq 1/n^{c+2}) \geq e^{-1/n^{c+1}} \geq 1 - 1/n^{c+1}$ for any fixed k and $\text{Prob}(X_{(n)} \geq 1/n^{c+2}) \geq 1 - 1/n^{c+1}$. All events together occur with probability $1 - O(1/n^c)$. By the same arguments as those in the proof of Theorem 7.9, the expected time until the discrepancy becomes at most t^* is $O(n^{c+4} \log n)$, and the time is bounded by $O(n^{c+4} \log^2 n)$ with probability $1 - O(1/n^c)$. The sum of all failure probabilities is $O(1/n^c)$. This proves the theorem.   □

Conclusions

In this chapter, we have studied $(1+1)$ EA_b and RLS_b^1 on an NP-hard scheduling problem. Two different perspectives were taken, namely regarding the worst-case and average-case models. In the worst case, the approximation ratios obtainable in polynomial time are bounded by roughly $4/3$. Using a result on the success probability and employing multistart variants of the search algorithms, we obtain a drastic improvement. The simple algorithms then serve as polynomial-time randomized approximation schemes. An even more encouraging result is obtained in the two average-case models investigated. With growing problem size, the makespan converges to optimality.

8

Shortest Paths

Computing shortest paths in a given graph is one of the fundamental problems in computer science. The input is given by a connected directed graph $G = (V, E)$ where $V = \{v_1, \ldots, v_n\}$ is a set of n vertices and E is a set of m edges. In addition, there is a weight function $w \colon E \to \mathbb{N}$ which assigns positive integer weights to the edges. We denote by $w_{\max} = \max_{e \in E} w(e)$ the maximum of the weights of all edges and distinguish between two problems. In the single-source shortest-path (SSSP) problem, there is one designated vertex $s \in V$ and the task is to compute a shortest path from s to every other vertex $v_i \in V \setminus \{s\}$. W. l. o. g., we assume $s = v_1$ throughout this chapter. The length of a path is measured by the sum of the weights of the edges that are used in this path. A generalization of the SSSP problem is the all-pairs shortest-path (APSP) problem, where the task is to compute from each vertex $v_i \in V$ a shortest path to every other vertex $v_j \in V \setminus \{v_i\}$. The SSSP and APSP problems can be solved by using Dijkstra's algorithm and the Floyd-Warshall algorithm, respectively. Using appropriate data structures, single-source shortest paths and all-pairs shortest paths can be computed in time $O(m + n \log n)$ and $O(nm + n^2 \log n)$, respectively (Mehlhorn and Sanders, 2008).

The basic algorithms for computing shortest paths in a given graph date back to the 1950s. However, the computation of shortest paths is still an important field of research. This especially holds for planning tasks in road networks where additional properties of the network can be taken into account (Bast, Funke, Sanders, and Schultes, 2007; Sanders and Schultes, 2006). On the other hand, several related problems in the area of routing and planning are NP-hard, and stochastic search algorithms have found many applications in this area. Examples are vehicle routing (El-Fallahi, Prins, and Calvo, 2008; Rizzoli, Montemanni, Lucibello, and Gambardella, 2007) and routing problems in computer networks (Dorigo and Stützle, 2004; Farooq, 2008; Kim and Choi, 2007). Therefore, it seems to be important to understand the basic problem of computing shortest paths in the context of stochastic search algorithms from a theoretical point of view in order to gain new insights that can help practitioners solve related problems arising in applications.

F. Neumann, C. Witt, *Bioinspired Computation*
in Combinatorial Optimization, Natural Computing Series,
DOI 10.1007/978-3-642-16544-3_8, © Springer-Verlag Berlin Heidelberg 2010

The chapter is organized as follows. We start by analyzing evolutionary algorithms for the computation of shortest paths and study in Section 8.1 how a variant of the (1+1) EA can solve the SSSP problem. For the APSP problem, we examine a population-based approach which computes for each pair of vertices a shortest path between them. The results presented in Section 8.2 show that using crossover and mutation as variation operators leads provably to faster evolutionary algorithms than algorithms that rely only on mutation. Having presented different results for evolutionary algorithms, we turn in Section 8.3 to ant colony optimization and point out that this kind of stochastic search algorithm leads even to better runtime bounds than the ones presented for evolutionary algorithms.

8.1 Single Source Shortest Paths

In the following, we discuss evolutionary algorithms for the SSSP problem. We will examine two approaches that rely on the fact that an optimal solution of the SSSP problem can be represented by a shortest path tree.

8.1.1 Evolutionary Algorithms

The SSSP problem consists of finding for each vertex $v_i \in V \setminus \{s\}$ a path from s to v_i. Scharnow, Tinnefeld, and Wegener (2004) have examined a representation of possible solutions where for each vertex v_i its predecessor $p(v_i)$ is stored. In this way directed graphs containing exactly $n - 1$ edges are represented. An optimal solution is a shortest path tree which contains for each vertex $v_i \in V \setminus \{s\}$ a shortest path from s to v_i. Note that we fixed $s = v_1$. Therefore, the search space consists of all $I = (p(v_2), \ldots, p(v_{n-1})) \in \{v_1, \ldots, v_n\}^{n-1}$ where $p(v_i) \neq v_i$. A search point I is a vector of length $n - 1$ which stores for each vertex $v_i \in V \setminus \{s\}$ its predecessor.

A mutation carries out a set of local operations. A local operation for the SSSP problem picks one vertex $v_i \neq s$ uniformly at random and replaces the predecessor $p(v_i)$ of v_i with another predecessor $p'(v_i) \in V \setminus \{v_i, p(v_i)\}$. For a mutation step, we choose S from a Poisson distribution with parameter $\lambda = 1$ and perform sequentially $S+1$ local operations. The (1+1) EA$_{SP}$ which we will analyze for the SSSP problem is given in Algorithm 11. The algorithm uses a fitness function f to determine whether to replace the current individual with the offspring.

We will investigate two different approaches for defining the fitness of a search point I. Consider a candidate solution $I = (p(v_2), \ldots, p(v_{n-1}))$. Associated with I consider the subgraph T_I of the input graph G consisting of those pairs $(v_j, p(v_j))$ which are edges in G. If there is a path in T_I from the source s to v_j, $s \neq v_j$, it has to be unique. Let $\gamma(v_j)$ denote the unique path in such cases. Whenever such a unique path $\gamma(v_j)$ exists for a vertex v_j, we define its cost $f_j(I)$ to be the sum of the weights of the edges in $\gamma(v_j)$. On

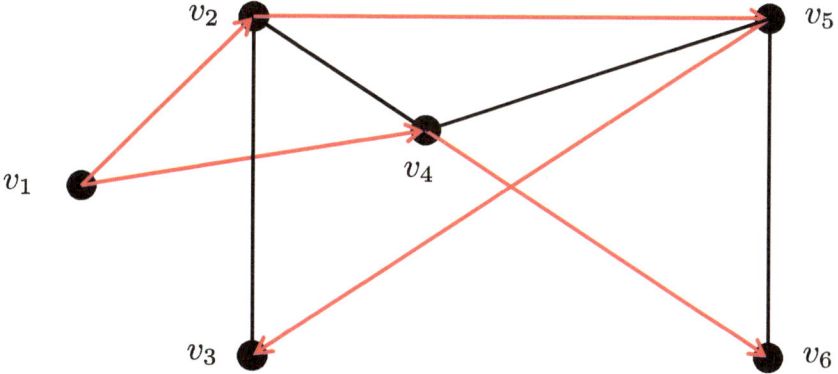

Fig. 8.1. Graph G marked with a tree given by the individual $I = (v_1, v_5, v_1, v_2, v_4)$

Algorithm 11 (1+1) EA$_{\mathrm{SP}}$

---

1. Set $I = (p(v_2), \dots p(v_n))$ where each $p(v_i) \in V \setminus \{v_i\}$ is chosen uniformly at random.
2. Choose S from a Poisson distribution with parameter $\lambda = 1$ and perform $S + 1$ local changes chosen uniformly at random to produce I'.
3. Replace I with I' if $f(I') \leq f(I)$.
4. Repeat Steps 2 and 3 forever.

---

the other hand, if v_j is unreachable from s in T_I, then the cost $f_j(I)$ is set to a large penalty value, i.e., $f_j(I) = d_{\mathrm{penalty}} := n \cdot w_{\max}$. First, we consider a multi-criteria fitness function which assigns to each search point a vector consisting of $n - 1$ components. In this vector, the length of the path from s to v_i is stored for each vertex $v_i \neq s$. Later on, we investigate a single-criterion fitness function which takes the sum of the lengths of the different paths into account.

8.1.2 Multi-Criteria Fitness Function

We consider the multi-criteria fitness function f_{mult} defined as

$$f_{\mathrm{mult}}(I) = (f_2(I), f_3(I), \dots f_n(I)).$$

We define $f_{\mathrm{mult}}(I') \leq f_{\mathrm{mult}}(I)$ iff $f_i(I') \leq f_i(I)$, $2 \leq i \leq n$. Using f_{mult} in the (1+1)EA$_{SP}$ we will see that the algorithm is able to follow the ideas of Dijkstra's algorithm for the computation of shortest paths. This leads to a polynomial bound on the expected optimization time. The basic result can be found in Scharnow et al. (2004) and the refined analysis which we present in the following is due Doerr, Happ, and Klein (2007a). To show an upper bound on the optimization time, we take the depth of the shortest path tree into account. We introduce the *edge radius* of the vertex s in the graph G, which is given by the following definition.

Definition 8.1. *The edge radius $\ell_G(s)$ of a vertex $s \in V$ of a given weighted graph G is the maximum number of edges in any shortest path with the minimum number of edges from s to v, i.e.,*

$$\ell_G(s) = \max_{v \in V}\{\min_{\gamma \in \Gamma_v} |\gamma|\}$$

where $\Gamma_v := \{\gamma \mid \gamma$ is a shortest path from s to $v\}$ and $|\gamma|$ denotes the number of edges in γ.

We use the abbreviation $\ell = \ell_G(s)$ and take this parameter into account for analyzing the expected optimization time.

Theorem 8.2. *The expected optimization time of (1+1) EA_{SP} using the fitness function f_{mult} is $O(n^2(\log n + \ell))$.*

Proof. We distinguish between two cases depending on the parameter ℓ. First, we investigate the case $\ell \geq \log n$ and show an upper bound of $O(n^2\ell)$. Consider a vertex $v \neq s$ and fix a shortest path $\gamma := (s = v^1, v^2, \ldots, v = v^{\ell'+1})$ where $\ell' \leq \ell$. Note that such a path exists according to Definition 8.1. There may be different possible shortest paths from s to v and optimal sub-paths of γ may be exchanged for different optimal sub-paths. However, this can only speed up the optimization process.

As γ is a shortest path from s to v, the sub-path $\gamma' = (s = v^1, v^2, \ldots, v^j)$ is a shortest path from s to v^j. We investigate a typical run and consider a phase of length $cn^2\ell$ where c is an appropriate constant. We assume that the current search point I already contains shortest paths from s to $v^2, \ldots, v^j$, $j < \ell'+1$. The local operation which sets $p(v^{j+1}) = v^j$ happens with probability at least $1/(en^2)$ in the next step. It produces from I a search point I' which contains shortest paths from s to $v^2, \ldots, v^j, v^{j+1}$. We call such an operation a success. A success happens in the next iteration with probability at least $p = 1/(en^2)$ independently of previous steps if a shortest path from s to v has not been obtained. If a shortest path from s to v has already been computed, we consider one fixed mutation operation that happens with probability at least $p = 1/(en^2)$ and define it as a success. Hence, a success happens in each iteration with probability at least $p = 1/(en^2)$ and we investigate the number of successes. Note that a shortest path from s to v has been obtained if the number of successes is at least ℓ.

When considering a phase of length $cn^2\ell$, the expected number of successes is $cn^2\ell/(en^2) = c\ell/e$. Using Chernoff bounds the probability that the number of successes is less than $(1-\delta)c\ell/e$ is upper bounded by $e^{-c\ell\delta^2/(2e)}$. Choosing $\delta = 1 - e/c$, the probability of having less than

$$(1 - \delta)c\ell/e = (1 - (1 - e/c))c\ell/e = \ell$$

successes is upper bounded by

$$e^{-c\ell(1-e/c)^2/(2e)} = e^{-c'\ell},$$

where $c' = c(1 - e/c)^2/(2e)$. Our previous computation holds independently for each vertex $v \neq s$. There are $n - 1$ vertices for which a shortest path has to be computed. The probability that a shortest path has not been computed for at least one vertex $v \neq s$ is therefore upper bounded by

$$(n - 1)e^{-c'\ell}.$$

Remember that we work under the assumption that $\ell \geq \log n$ holds. Hence, within a phase of $cn^2\ell$ steps all shortest paths have been computed with probability at least

$$\alpha = 1 - (n - 1)e^{-c'\ell} = 1 - O(n^{1-c'}).$$

The expected number of phases of lenth $cn^2\ell$ is upper bounded upper by α^{-1} which leads to an upper bound of $\alpha^{-1}cn^2\ell = O(n^2\ell)$ on the expected time to compute an optimal solution.

In the case where $\ell < \log n$ holds, we consider a phase of $cn^2 \log n$ steps, c again an appropriate constant. Following the ideas of the first case, an optimal solution has been obtained with probability at least

$$\alpha = 1 - (n - 1)e^{-c' \log n} = 1 - O(n^{1-c'})$$

and an optimal solution has been found after an expected number of at most $\alpha^{-1}cn^2 \log n = O(n^2 \log n)$ steps. This completes the proof.   □

The previous theorem shows that (1+1) EA$_{\mathrm{SP}}$ solves the SSSP efficiently when using the multi-criteria fitness function. A basic property when using this fitness function is that shortest paths that have been obtained during the optimization process cannot get lost. This might happen when considering a single-criterion fitness function which takes the sum of the different path lengths into account. We want to examine such an approach in the next section.

8.1.3 Single-Criterion Fitness Function

Using penalty values is a common approach for handling constraints (Michalewicz, 1995) and leads the algorithm towards feasible solutions. When large penalty values are used for vertices v that are not connected to the source, it does not seem necessary to use a multi-criteria fitness function.

Therefore, we investigate a single-objective fitness function which leads the algorithm towards valid solutions. The fitness of a candidate solution I is given by

$$f_{\text{sing}}(I) := \sum_{i=2}^{n} f_i(I),$$

which returns the sum of the different path lengths. Note that vertices not connected to s contribute a large value $d_{\text{penalty}} = n \cdot w_{\max}$ to the fitness value.

This kind of fitness function has already been proposed in Scharnow et al. (2004). However, it took quite some time until Baswana, Biswas, Doerr, Friedrich, Kurur, and Neumann (2009) were able to show that it leads to a polynomial upper bound on the expected optimization time of (1+1) EA$_{\text{SP}}$. The idea is to show that there is always a set of local operations which reduces the difference between the fitness of the current solution and an optimal one by the fraction $1/n$. This enables us to use the method of the expected multiplicative distance decrease and leads to the following result.

Theorem 8.3. *The expected optimization time of (1+1) EA$_{SP}$ using the fitness function f_{sing} is $O(n^3 \cdot (\log n + \log w_{\max}))$.*

Proof. Let I_{opt} be an optimal search point and T_{opt} be the corresponding shortest path tree. We define the total distance of the current solution I from an optimal one as

$$d = f_{\text{sing}}(I) - f_{\text{sing}}(I_{\text{opt}}).$$

The total distance can be split into $d(v^i) = f_i(I) - f_i(I_{\text{opt}})$, $2 \le i \le n$. We define $d(v^1) = 0$. Note that $d = \sum_{i=2}^{n} d(v^i)$. Hence, there is at least one vertex v with distance at least d/n. Consider the path $\gamma = (s = v^1, v^2, \ldots, v = v^{\ell'})$ from s to v in T_{opt}. Obviously,

$$d(v) = d(v^{\ell'}) - d(v^1) = \sum_{j=2}^{\ell'} d(v^j) - d(v^{j-1}) \ge d/n$$

holds.

Consider two vertices v^i and v^{i+1} in γ for which $d(v^i) < d(v^{i+1})$ holds. At such a pair of vertices the distance increases and we call the edge (v^i, v^{i+1}) positive. Note that if (v^i, v^{i+1}) is a positive edge, v^i has to be connected to the source as otherwise $d(v^{i+1}) \le d(v^i)$. On the other hand, $p(v^{i+1}) \ne v^i$ holds as otherwise (v^i, v^{i+1}) is not a positive edge. Setting $p(v^{i+1}) = v^i$ implies that $d(v^{i+1}) = d(v^i)$. Hence, such an operation is accepted and reduces the distance by $d(v^{i+1}) - d(v^i)$. Considering all positive edges in γ, we can achieve $d(v) = 0$ by setting $p(v^{i+1}) = v^i$ for each positive edge (v^i, v^{i+1}) in γ. Each of these operations is accepted and the total distance decrease is at least d/n. Denote by k, $1 \le k \le n-1$, the number of positive edges in γ. We may add $n-k$ non-accepted operations changing the predecessor of a particular vertex such that the total number of considered operations is n. The probability of choosing one of these operations in the next step is $\Omega(n/n^2) = \Omega(1/n)$ and the average distance decrease of these n operations is at least d/n^2. Hence, the expected distance after such an operation has happened is $(1 - 1/n^2) \cdot d$. Following the ideas of the expected multiplicative distance decrease (see Section 4.2.3) and taking into account that the distance of the initial solution is upper bounded by $d_{\max} = (n-1)d_{\text{penalty}} = (n-1) \cdot n \cdot w_{\max}$, the expected optimization time is upper bounded by $O\left(n^3 \cdot (\log n + \log w_{\max})\right)$ □

8.2 All Pairs Shortest Paths

Having examined how evolutionary algorithms can cope with the SSSP problem we turn to the APSP problem. For the APSP problem we examine population-based evolutionary algorithms. Each individual of the population P is a path. Our goal is to evolve an initial population consisting of a set of paths into a population which contains for each pair of vertices (u, v), $u \neq v$, a shortest path from u to v. The approach and the results that we present in this section are due to Doerr, Happ, and Klein (2008).

We investigate two evolutionary algorithms for the APSP problem. The first one, called Steady State EA$_{\text{SP}}$ (see Algorithm 12), works with mutation as a variation operator. Our second algorithm, called Steady State GA$_{\text{SP}}$ (see Algorithm 13), relies on crossover and mutation. Both algorithms start with a population $P := \{I_{u,v} = (u, v) | (u, v) \in E\}$ of size $|E|$ which contains all paths corresponding to the edges of the given graph G. The variation operators produce in each iteration one single offspring. In Steady State EA$_{\text{SP}}$ an offspring is obtained by choosing one individual uniformly at random from the population and applying a mutation operator. In Steady State GA$_{\text{SP}}$, either with probability p_c a crossover operator is applied to two randomly chosen individuals of P or (if this is not the case) mutation is used as in Steady State EA$_{\text{SP}}$. Note, that both algorithms are equivalent if $p_c = 0$. In the following, we assume that $0 < p_c < 1$ is a constant. The selection operator only accepts individuals that are paths in the graph. In addition, it ensures diversity with respect to the different pairs of vertices. Each individual $I_{u,v}$ that is a valid path is indexed by the start vertex u and the end vertex v. In the selection step an offspring is only compared to an individual of the current population that has the same start and end vertex. It is ensured that for each pair of vertices (u, v), $u \neq v$, at most one individual $I_{u,v}$ is contained in the population. This implies that the population size of our algorithms is always at most $n(n-1)$.

The mutation operator takes an individual $I_{u,v}$ from the population and applies sequentially $S + 1$ local operations. Here, S is a parameter that is chosen according to the Poisson distribution with parameter $\lambda = 1$. In a local operation, the current path is either lengthened or shortened by a single edge. Assume that the current individual represents a path $\gamma = (u = v^1, v_2, \ldots v^{\ell'}, v = v^{\ell'+1})$ from u to v consisting of ℓ' edges, and denote by $E^-(v)$ and $E^+(v)$ the set of incoming and outgoing edges of a vertex v in G. Then an edge

$$e = (x, y) \in E^-(u) \cup E^+(v) \cup \{(u, v^2), (v^{\ell'}, v)\}$$

Algorithm 12 Steady State EA$_{\mathrm{SP}}$
---

1. Set $P = \{I_{u,v} = (u, v) \mid (u, v) \in E\}$.
2. Choose an individual $I_{x,y} \in P$ uniformly at random.
3. Mutate $I_{x,y}$ to obtain an individual $I'_{s,t}$.
4. If there is no individual $I_{s,t} \in P$, $P = P \cup \{I'_{s,t}\}$,
   else if $f(I'_{s,t}) \leq f(I_{s,t})$, $P = (P \cup \{I'_{s,t}\}) \setminus \{I_{s,t}\}$
5. Repeat Steps 2–4 forever.

Algorithm 13 Steady State GA$_{\mathrm{SP}}$
---

1. Set $P = \{I_{u,v} = (u, v) \mid (u, v) \in E\}$.
2. Choose $r \in [0, 1]$ uniformly at random.
3. If $r \leq p_c$, choose two individuals $I_{x,y} \in P$ and $I_{x',y'} \in P$ uniformly at random
   and perform crossover to obtain an individual $I'_{s,t}$, ·
   else choose an individual $I_{x,y} \in P$ uniformly at random and mutate $I_{x,y}$ to
   obtain an individual $I'_{s,t}$.
4. If $I'_{s,t}$ is a path from s to t then
   a) If there is no individual $I_{s,t} \in P$, $P = P \cup \{I'_{s,t}\}$,
   b) else if $f(I'_{s,t}) \leq f(I_{s,t})$, $P = (P \cup \{I'_{s,t}\}) \setminus \{I_{s,t}\}$.
5. Repeat Steps 2–4 forever.

---

is chosen uniformly at random. If

$$e \in \{(u, v^2), (v^{\ell'}, v)\},$$

the edge is removed. This means that either the first edge or the last edge in the path is removed, leading to an individual $I'_{v^2,v}$ or $I'_{u,v^{\ell'}}$ consisting of $\ell' - 1$ edges.
   If

$$e \in (E^-(u) \cup E^+(v)) \setminus \{(u, v^2), (v^{\ell'}, v)\},$$

the edge is added and the path is lengthened. Here, a new individual $I'_{x,v}$ or $I'_{u,y}$ is produced that contains $\ell' + 1$ edges. Note that a local operation applied to a valid path always leads to a new valid solution, which implies that the mutation operator only constructs solutions which are paths.

   In Doerr et al. (2008), different crossover operators have been discussed which are all motivated by the 1-point crossover operator known from binary encodings. We discuss one of them and note that the other operators lead to the same runtime bounds. Our crossover operator chooses two individuals $I_{u,v}$ and $I_{w,x}$ uniformly at random from the population and produces a new individual $I'_{u,x}$ if $v = w$ holds. Note that the crossover operator only constructs a valid solution, namely a path from u to x, iff the end vertex of the first individual equals the start vertex of the second individual.

First, we will consider how Steady State EA$_{\text{SP}}$ can solve the APSP problem. Later on, we will show that the use of crossover leads to improved runtime bounds.

8.2.1 Results for Steady State EA$_{\text{SP}}$

In the following, we consider the algorithm that uses only mutation as variation operator. We already know that the (1+1) EA$_{\text{SP}}$ introduced in Section 8.1 computes a shortest path from a predefined vertex s to any other vertex $v \neq s$ in the given graph in time $O(n^3)$. Therefore, the APSP problem can be solved by applying the (1+1) EA$_{\text{SP}}$ for each given vertex in the graph. This leads to n runs of the algorithm which can be carried out sequentially and leads to a solution for the APSP problem by evolutionary algorithms in expected time $O(n^4)$. Note that these n runs of the (1+1) EA$_{\text{SP}}$ can also be carried out in parallel.

We now study Steady State EA$_{\text{SP}}$ and show that this algorithm produces in expected time $O(n^4)$ a population which contains for any two vertices u and v an individual representing a shortest path from u to v. The actual upper bound again depends on the edge radius $\ell := \ell(G)$ defined in Definition 8.1.

Lemma 8.4. *Let $\ell \geq \log n$. The expected time until Steady State EA$_{SP}$ has found all shortest paths with at most ℓ edges is $O(n^3 \ell)$.*

Proof. To prove the lemma, we can reuse the ideas used in the proof of Theorem 8.2. In this proof an upper bound of $O(n^2 \ell)$ has been shown for the case where $\ell \geq \log n$ holds. The major difference is that Steady State EA$_{\text{SP}}$ works with a population whose size is bounded by $O(n^2)$.

Consider two vertices u and v, $u \neq v$, and let $\gamma := (v^1 = u, v^2, \ldots, v^{\ell'+1} = v)$ be a shortest path from u to v consisting of ℓ', $\ell' \leq \ell$, edges in G. As γ is a shortest path from u to v, the sub-path $\gamma' = (v^1 = u, v^2, \ldots, v^j)$ is a shortest path from u to v^j. Again, we investigate a typical run but consider this time a phase of length $cn^3 \ell$ where c is an appropriate constant. We assume that the current population already contains individuals that represent shortest paths from u to $v^2, \ldots, v^j$, $j < \ell' + 1$.

Then there is a local operation which picks the individual representing the shortest path from u to v^j and produces an individual representing a shortest path from u to v^{j+1}. The probability that such a step happens in the next iteration is at least $1/(2n^3)$ as the population size is bounded by n^2 and the probability of appending the right edge to the shortest path from u to v^j is at least $1/(2en)$. We call such an operation a success. A success happens in the next iteration with probability at least $p = 1/(2en^3)$ independently of previous steps. If a shortest path from u to v has already been computed, we consider one fixed mutation operation that happens with probability at least $p = 1/(2en^3)$ and define it as a success. Hence, a success happens in each iteration with probability at least $p = 1/(2en^3)$. We may lower bound the probability of having for each pair of vertices enough successes by $\alpha = 1 - o(1)$ using similar calculations as in the proof of Theorem 8.2. $\square$

As the number of edges in any shortest path is upper bounded by $n - 1$ we get the following theorem.

Theorem 8.5. *The expected optimization time of Steady State* EA_{SP} *is* $O(n^4)$.

Doerr et al. (2008) have investigated a worst-case example for Steady State EA_{SP}. The example is the complete directed graph where all edges in the path $(v_1, v_2, \ldots, v_n)$ get weight 1 and all other edges get a large weight of n. For this input graph they have shown a lower bound of $\Omega(n^4)$ on the expected optimization time.

8.2.2 Results for Steady State GA$_{SP}$

Our investigations for the Steady State EA_{SP} have shown that this algorithm computes a population representing for each pair of vertices a shortest path in expected time $O(n^4)$. Due to the lower bound for the complete directed graph given by Doerr et al. (2008), the question arises about whether the computation can be sped up by using a crossover operator. We will examine Steady State GA$_{SP}$ where the probability of using crossover is a constant. To make sure that both operators, mutation and crossover, are used we require $p_c \notin \{0, 1\}$. All the following results hold if the crossover probability is chosen as an arbitrary constant, i.e., $p_c \in \,]0, 1[$ and $p_c = \Omega(1)$.

Theorem 8.6. *The expected optimization time of Steady State* GA$_{SP}$ *is* $O(n^{3.5}\sqrt{\log n})$.

Proof. The main idea is that the mutation operator constructs for any pair of vertices for which a shortest path of at most $\ell^* := \sqrt{n \log n}$ edges exists such a solution. For pairs of vertices for which no shortest path of at most ℓ^* exists, the crossover operator constructs a shortest path by joining shortest paths of a smaller number of edges.

Mutation is used with probability $1 - p_c = \Omega(1)$ in each iteration. Hence, all shortest paths with at most ℓ^* edges are obtained in expected time $O(n^3 \ell^*) = O(n^3 \sqrt{n \log n}) = O(n^{3.5} \log n)$ due to Lemma 8.4.

In the following, we work under the assumption that all shortest paths with at most ℓ^* edges have already been obtained and examine how to obtain shortest paths consisting of more than ℓ^* edges by crossover.

We assume that the population contains an individual $I_{u,v}$ which represents a shortest path from u to v if there exists such a path containing at most k edges. We consider a pair of vertices x and y for which a shortest path of at most r, $k < r \leq 2k$, edges exists. The shortest path from x to y of length r can be split up at $2k - r$ positions such that two paths of lengths at most k are obtained. Hence, there are $2k - r$ pairs of paths in the population which can be joined to obtain the desired path of length r. Each of these pairs of paths is selected with probability at least $1/n^4$ in the next step as the population size is upper bounded by n^2. Taking the number of different pairs into account, the probability of selecting two paths from the population which are joined by the crossover operator to the desired shortest path from x to y is $\Omega(\frac{2k+1-r}{n^4})$. Note that this probability is $\Omega(\frac{k}{n^4})$ for $r \leq \frac{3k}{2}$. There are at most n^2 paths of at most r edges. Assuming that all shortest paths consisting of at most k edges are already contained in the population, the expected number of additional steps until all shortest paths containing at most $\frac{3k}{2}$ are in the population is $O(\frac{n^4 \log n}{k})$ using arguments from the coupon collectors problem. The upper bound on the expected optimization time can be computed by summing up over the different values of k, namely

$$\sqrt{n \log n}, c \cdot \sqrt{n \log n}, c^2 \cdot \sqrt{n \log n}, \ldots, c^{\log_c(n/\sqrt{n \log n})} \cdot \sqrt{n \log n},$$

where $c = 3/2$. Hence, the expected optimization time is upper bounded by

$$\sum_{s=0}^{\log_c(n/\sqrt{n \log n})} \left(O\left(\frac{n^4 \log n}{\sqrt{n \log n}} \right) c^{-s} \right) = O(n^{3.5}\sqrt{\log n}) \sum_{s=0}^{\infty} c^{-s} = O(n^{3.5}\sqrt{\log n})$$

□

Using a more sophisticated analysis, the upper bound on the expected optimization time of Steady State GA$_{SP}$ may be improved to $O(n^{3.25} \log^{1/4} n)$. We refer the interested reader to the original work of Doerr and Theile (2009). In this work, also a worst-case graph is given and a lower bound of $\Omega(n^{3.25} \log^{1/4} n)$ on the expected optimization time of Steady State GA$_{SP}$ is proven.

8.3 Analysis of Ant Colony Optimization

In this section, we revisit the ACO framework that was introduced in Section 5.4. Both the SSSP and APSP problems are considered, based on studies by Attiratanasunthron and Fakcharoenphol (2008) and Horoba and Sudholt (2009). Shortest path problems are maybe the most natural combinatorial optimization problems to be treated by ACO since these search algorithms were inspired by the way ants find shortest paths to food sources.

Both the SSSP and the APSP problems make sense in the simple case of undirected, weighted graphs. ACO algorithms, however, implicitly direct edges according to the direction they are being traversed by the artificial ants. Since it turns out to be more convenient, we replace the SSSP problem in this section with the *single-destination shortest path* (SDSP) problem, where the aim is to find shortest paths from every vertex to a specified destination vertex. Given a directed graph, the two problem variants can be converted into each other just by turning around the directions of all edges. Hence, the SSSP and SDSP problem are conceptually equivalent, but the consideration of a destination will ease the presentation of the ACO framework.

Throughout this section, we denote by d the destination vertex of the SDSP problem. The ACO approach is population-based and uses $n = |V|$ ants that proceed, to some extent, independently. From each vertex $u \in V$, there is one ant a_u heading for the destination. The walk of each ant is again controlled by pheromone values $\tau \colon E \to \mathbb{R}^+$ that are global variables in the algorithm (so the same pheromone values control different ants). Heuristic information is not used. In order to complete the path construction in linear time, we disallow vertices from being visited more than once. This leads for ant a_u to the procedure described in Algorithm 14. Note that the destination d is not necessarily reached if the ant takes wrong decisions. In this case, we define the length of the path output by Algorithm 14 to be infinite.

Algorithm 14 Path construction from u to d

$i \leftarrow 0$.
$p_i \leftarrow u$.
$V_1 \leftarrow \{p \in V \setminus \{p_0\} \mid (p_0, p) \in E\}$.
while $p_i \neq d$ and $V_{i+1} \neq \emptyset$ **do**
$\quad i \leftarrow i + 1$.
$\quad$ Choose $p_i \in V_i$ with probability $\tau((p_{i-1}, p_i)) / \sum_{p \in V_i} \tau((p_{i-1}, p))$.
$\quad V_{i+1} \leftarrow \{p \in V \setminus \{p_0, \ldots, p_i\} \mid (p_i, p) \in E\}$.
end while
return $(p_0, \ldots, p_i)$.

While the ants walk through the graph to probabilistically construct short paths from their start vertex to d, each ant memorizes the best path it has found so far. Initially, all best-so-far paths are empty, which corresponds to infinite length. Pheromone initialization and update will be described below. The top-level framework of the ACO approach, called MMAS$_{\text{SDSP}}$, is displayed in Algorithm 15. Note that one iteration of the main loop corresponds to n constructed solutions and, therefore, n evaluations of the objective function.

Finally, pheromone update and initialization are similar as in Section 5.4. Since we want the first path to choose successors of a vertex uniformly, we initialize for each vertex $u \in V$ the pheromones on the outgoing edges $e = (u, \cdot)$ according to $\tau(e) = 1/\text{outdeg}(u)$, i.e., inversely proportional to the outdegree of the vertex. Vertices of outdegree 1 are special since we keep the value $\tau(e) = 1$ fixed on the unique outgoing edge e throughout the run of the algorithm. When pheromones are updated, the n best-so-far solutions $p_1^*, \ldots, p_n^*$ are consulted, where solution p_u^*, $1 \leq u \leq n$, is used to update the pheromones on the outgoing edges of u according to

---

Algorithm 15 MMAS$_{\text{SDSP}}$

Initialize pheromones τ and best-so-far paths $p_1^*, \ldots, p_n^*$.
loop
  for $u = 1$ to n **do**
    Construct a simple path $p_u = (p_{u,0}, \ldots, p_{u,\ell_u})$ from u to d with respect to τ.
    if $w(p_u) < w(p_u^*)$ **then**
      $p_u^* \leftarrow p_u$
    end if
  end for
  Update pheromones with respect to $p_1^*, \ldots, p_n^*$.
end loop

$$\tau(e) \leftarrow \begin{cases} \min\{(1 - \rho) \cdot \tau(e) + \rho,\ \tau_{\max}\} & \text{if } e = (u, v) \in p_u^*, \\ \max\{(1 - \rho) \cdot \tau(e) + \rho,\ \tau_{\min}\} & \text{if } e = (u, v) \notin p_u^*. \end{cases}$$

Here, $\tau_{\max}$ and $\tau_{\min}$ are again bounds on the pheromone values that are typical for the MMAS approach. We will consider different choices of these bounds but ensure in any case that $\tau_{\max} + \tau_{\min} = 1$.

8.3.1 Single-Destination Shortest Path

In this section, we show that MMAS$_{\text{SDSP}}$ finds the shortest path in polynomial time. The exact bound depends on the parameters, one of which was relevant in the analysis of evolutionary algorithms earlier in this chapter. Namely, we consider the maximum outdegree $\Delta(G) := \max_{v \in V} \text{outdeg}(v)$ and the edge radius $\ell(G)$ of the graph (see Definition 8.1). To show the forthcoming theorem, a careful inspection of pheromone values is necessary.

Lemma 8.7. *For every vertex u with* $\text{outdeg}(u) > 1$ *it holds that*

$$1 \leq \sum_{e=(u,\cdot)\in E} \tau(e) \leq 1 + \text{outdeg}(u) \cdot \tau_{\min}.$$

Proof. Initially the sum of pheromones on outgoing edges of u equals 1. Assume for induction that $\sum \tau(e) \geq 1$. If the pheromones are not capped by the bound $\tau_{\max}$ then $(1 - \rho) \sum \tau(e) + \rho \geq 1$ holds after the pheromone update. If at least one pheromone is capped at $\tau_{\max}$ then the sum of pheromones is still at least $\tau_{\max} + \tau_{\min} \geq 1$ as $\text{outdeg}(u) \geq 2$.

    For the second inequality, observe that the sum of pheromones can only increase if a pheromone value is maximized with the lower border $\tau_{\min}$. The reason is that $\sum \tau(e) \geq 1$ implies that $(1 - \rho) \sum \tau(e) + \rho \leq \sum \tau(e) \geq 1$. Consider an edge e with $(1 - \rho)\tau(e) < \tau_{\min}$. When its pheromone value is set to the lower border then the difference from the former value is at most $\tau_{\min} - \tau(e) + \rho \cdot \tau(e) \leq \tau_{\min} \cdot \rho$, where we used $\tau(e) \geq \tau_{\min}$ and $\rho \leq 1$. If currently $\sum \tau(e) \leq 1 + \text{outdeg}(u) \cdot \tau_{\min}$ then the sum of the pheromone values after the next update is at most $(1 - \rho)(1 + \text{outdeg}(u)\tau_{\min}) + \rho + \text{outdeg}(u)\tau_{\min} \cdot \rho = 1 + \text{outdeg}(u) \cdot \tau_{\min}$. Hence, the second inequality follows by induction. ☐

As an immediate consequence, we obtain the following direct relation between pheromone values and probabilities that ant a_u, i.e., the ant starting at u, chooses an edge $(u, \cdot)$. We suppose $\tau_{\min} \leq 1/\text{outdeg}(u)$ since $\tau_{\min}$ should be chosen below the initial pheromone value of $1/\text{outdeg}(u)$.

Corollary 8.8. *If $\tau_{\min} \leq 1/\text{outdeg}(u)$ for every edge $e = (u, \cdot)$ then*

$$\frac{\tau(e)}{2} \leq \text{Prob}(\text{ant } a_u \text{ chooses edge } e) \leq \tau(e).$$

The lower bound also holds for every other ant leaving vertex u and every edge $e = (u, v)$ unless v has already been traversed by the ant. The upper bound also holds for every other ant and every edge $e = (u, \cdot)$ unless the ant has travesed a successor of u before.

Proof. The upper bound holds since $\sum_{e'=(u,\cdot)\in E} \tau(e') \geq 1$ according to Lemma 8.7 and the probability of choosing edge e is proportional to $\tau(e)$. For the upper bound, we note that $\sum_{e'=(u,\cdot)\in E} \tau(e') \leq 1 + \text{outdeg}(u)\tau_{\min} \leq 2$ since $\tau_{\min} \leq 1/\text{outdeg}(u)$. If some successors of v have already been visited, this only increases the probability of visiting an unvisited neighbor. ☐

Moreover, we need a technical lemma to analyze the number of iterations that suffice to raise a pheromone value from its lower to its upper border and vice versa.

Lemma 8.9. *Suppose that edge e is rewarded in each iteration of $MMAS_{SDSP}$. Then $\tau(e) = \tau_{\max}$ holds after at most $T^* := \log(\tau_{\max}/\tau_{\min})/\rho$ iterations. If e is never rewarded then $\tau(e) = \tau_{\min}$ holds after also at most T^* iterations.*

Proof. Both statements are proved together by investigating the following symmetrical situation: Let e_1 be an edge with initial pheromone value $\tau_{\min} = 1 - \tau_{\max}$ and e_2 be an edge with initial value $\tau_{\max}$. Assume further that e_1 is rewarded in each iteration while e_2 is never rewarded. Obviously $\tau(e_1) + \tau(e_2) = 1$ holds in the beginning, and the sum of pheromone values remains 1 since for all following iterations

$$(1 - \rho)\tau(e_1) + ((1 - \rho)\tau(e_2) + \rho) = \tau(e_1) + \tau(e_2) - \rho(\tau(e_1) + \tau(e_2)) + \rho = 1.$$

Hence, the time $\tau(e_1)$ reaches $\tau_{\max}$ equals the time $\tau(e_2)$ reaches $\tau_{\min}$. We therefore only study the time for the latter.

Since e_2 is never rewarded, it holds that $\tau(e_2) = (1 - \rho)^t \cdot \tau_{\max}$ after t iterations or the lower pheromone border is reached. Solving the equation

$$(1 - \rho)^T \cdot \tau_{\max} \leq \tau_{\min}$$

with respect to T yields $T \leq \ln(\tau_{\min}/\tau_{\max})/\ln(1 - \rho)$ which, using $\ln(1 - \rho) \leq -\rho$, implies the lemma. ☐

The following theorem gives two upper bounds for MMAS$_{\text{SDSP}}$, each consisting of two additive terms. Intuitively, the first terms cover waiting times until improvements of best-so-far paths have been found. The second terms grow with $1/\rho$. They reflect the time to adapt the pheromones after a change of the best-so-far path. This time is called *freezing time* by Neumann et al. (2009). The theorem is restricted to acyclic graphs but allows (other than the results from the beginning of this chapter) negative weights. A statement for graphs that may contain cycles but no negative weights is given afterwards.

Theorem 8.10. *Consider a directed acyclic graph G. The expected number of iterations of MMAS$_{SDSP}$ on G with $\tau_{\min} := 1/n^2$ is $O(n^3 + (n \log n)/\rho)$. Let $\Delta := \Delta(G)$ and $\ell = \ell(G)$. Choosing $\tau_{\min} = 1/(\Delta \ell)$ the expected number of iterations is $O(n\Delta\ell + n\log(\Delta\ell)/\rho)$.*

Proof. We start by defining the following notions. Call an edge (u, v) *incorrect* if it does not belong to any shortest path from u to d. We say that a vertex u has been *processed* if a shortest path from u to d has been found and if all incorrect edges leaving u have pheromone $\tau_{\min}$.

The proof proceeds inductively. We estimate the expected time (where time is measured in iterations of the main loop of MMAS$_{\text{SDSP}}$) until a vertex u has been processed given that all vertices reachable from u on shortest paths from u to d have already been processed. To this end, we first consider the expected time until a shortest path from u to d has been found for the first time. We say then that vertex u has been optimized. By Corollary 8.8, the probability of choosing an outgoing edge of u that belongs to a shortest path from u to d is at least $\tau_{\min}/2$. Since all vertices reachable from u have been processed, all incorrect edges at any reachable vertex v have pheromone $\tau_{\min}$ and the probability of choosing some incorrect edge is at most outdeg$(v)\tau_{\min}$. Hence, the probability of continuing in the construction procedure on an edge on a shortest path is at least $1 - \text{outdeg}(v)\tau_{\min} \geq 1 - 1/\ell$ if $\tau_{\min} \leq 1/(\Delta\ell)$. As there is, by definition, a shortest path with at most ℓ edges, the probability that no incorrect edge is chosen on the way from v to d is at least $(1-1/\ell)^{\ell-1} \geq 1/e$. Together with the choice of an appropriate successor of u, the probability of optimizing u is at least $\tau_{\min}/(2e)$.

The expected time until u is optimized is thus at most $2e/\tau_{\min}$. Afterwards, the update mechanism of MMAS$_{\text{SDSP}}$ ensures that a shortest path from u to d is reinforced automatically in each iteration. The precise path may change, but it is guaranteed that only shortest paths are rewarded and hence the pheromone on incorrect edges decreases in every iteration. Lemma 8.9 states that $\ln(\tau_{\max}/\tau_{\min})/\rho$ iterations are enough for the vertex to become processed; hence the expected time until u is processed is at most $2e/\tau_{\min} + \ln(\tau_{\max}/\tau_{\min})/\rho$.

Finally, the inductive argument is made precise. Let $v^0 = d, v^1, \ldots, v^{n-1}$ be a topological ordering of the vertices starting from the destination d, i.e., for every edge $(u, v) \in E$ it holds that v precedes u in the ordering. Such a topological ordering exists since G is assumed to be acyclic. Consequently, all shortest paths from v^i to d only use vertices from $\{v^0, \ldots, v^{i-1}\}$. If $v^1, \ldots, v^{i-1}$ have been processed then we can wait for v^i to be processed using the above argumentation. The expected time until all vertices $v^1, \ldots, v^{n-1}$ have been processed is bounded by $2en/\tau_{\min} + n\ln(\tau_{\max}/\tau_{\min})/\rho$. Choosing $\tau_{\min} = 1/n^2$, we obtain the bound $O(n^3 + (n\log n)/\rho)$, and choosing $\tau_{\min} = 1/(\Delta\ell)$, we obtain $O(n\Delta\ell + n\log(\Delta\ell)/\rho)$. $\square$

The bound from the previous theorem can be improved by a factor of ℓ/n if the shortest paths are unique and ℓ is not too small. Moreover, we can drop the assumption of acyclic graphs and again demand positive weights instead.

Theorem 8.11. *Consider a directed graph G with positive weights where all shortest paths are unique. If $\ell := \ell(G) \geq \ln n$ holds, the number of iterations of MMAS$_{SDSP}$ on G with $\tau_{\min} = 1/(\Delta\ell)$ is bounded from above by $O(\Delta\ell^2 + \ell\log(\Delta\ell)/\rho)$ with probability at least $1 - 1/n^2$. The bound on the number of iterations holds also in expectation.*

Proof. The main change from the proof of Theorem 8.10 is that here the vertices to be processed are enumerated in a different order. Since all shortest paths are unique, they all have length at most ℓ, and we wait for MMAS$_{SDSP}$ to process every shortest path in reverse order. Doing this, we exploit the fact that all weights are positive and additionally use a concentration result similar to the one from the proof of Theorem 8.2.

Let u be an arbitrary but fixed vertex and let $u = v^{\ell'}, v^{\ell'-1}, \ldots, v^0 = d$ be the unique shortest path of length $\ell' \leq \ell$ from u to d. Since all weights are positive, all shortest paths from v^i to d, $1 \leq i \leq \ell'$, use only vertices from $\{v^0, \ldots, v^{i-1}\}$. Hence, if $v^1, \ldots, v^{i-1}$ have been processed then we can wait for v^i has been processed. Since $\tau_{\min} = 1/(\Delta\ell)$, the probability of finding a shortest path from u to d given the processed successors is at least $\tau_{\min}/(2e)$ using the same argumentation as that in the proof of Theorem 8.10.

Given processed successors, let T_i denote the random time (number of iterations) until v^i is optimized (recall that this notion does not yet imply v^i to be processed). Consider random variables $X_1, \ldots, X_T$ that are independently set to 1 with probability $\tau_{\min}/(2e)$ and to 0 otherwise. The random first point of time T_1^* where $X_1 = 1$ stochastically dominates the random time until v^1 is optimized. As v^1 becomes processed after an additional (deterministic) waiting time of at most $F := \ln(\tau_{\max}/\tau_{\min})/\rho$ iterations, $T_1^* + F$ stochastically dominates T_1. Inductively, we have that $T_{\ell'}^* + \ell'F$ stochastically dominates $T_{\ell'}$ and hence the time until u is processed.

Let $T := 16e\ell/\tau_{\min}$ and $X := \sum_{i=1}^{T}$. We have $E(X) = T \cdot \tau_{\min}/(2e) = 8\ell$. By Chernoff bounds,

$$\text{Prob}(X < \ell) \leq \text{Prob}(X \leq (1 - 7/8) \cdot E(X)) \leq e^{-8\ell^*(7/8)^2/2} < e^{-3\ell^*} \leq 1/n^3,$$

where we used the assumption $\ell \geq \ln n$. Hence, the probability that u is not processed after $T + \ell F = O(\Delta\ell^2 + \ell\log(\Delta\ell)/\rho)$ iterations is at most $1/n^3$. By the union bound, the probability that there is an unprocessed vertex remaining after this time is at most $1/n^2$. This proves the first statement of the theorem.

The preceding argumentation holds for arbitrary initilization of the ACO algorithm, in particular, if the pheromone values have assumed an arbitrary value. Hence, we can repeat the argumentation with another phase of $T + \ell F$ iterations if the algorithm does not find all shortest paths within the first $T + \ell F$ iterations. The expected number of phases is $1 + o(1)$, which implies the bound on the expected number of iterations.   □

Horoba and Sudholt (2009) also show a lower bound that is tight with the upper bound from Theorem 8.11 if $\rho = \Omega((\log n)/n)$, and almost tight with a gap of at most $O(\log^2 n)$ otherwise. The underlying graph is similar to the example mentioned above following Theorem 8.5. All vertices are lined up on a single path that contains all shortest paths as subsets. Then MMAS$_{\mathrm{SDSP}}$ has to optimize $n - 1$ vertices sequentially with respect to increasing distance from the destination.

The lower bound shows that no significant improvements are possible with the setting of Theorem 8.11. However, there is room for improvement if the degrees of the vertices differ significantly. Rather than choosing $\tau_{\min} = 1/(\Delta\ell)$ for all vertices, we set the pheromone bounds of edges adaptively with respect to the vertices they are leaving. More precisely, for every vertex u, we restrict the pheromone values of the outgoing edges $e = (u, \cdot)$ to the interval $[\tau_{\min}(u), \tau_{\max}(u) = 1 - \tau_{\min}(u)]$. The choice $\tau_{\min}(u) := 1/(\mathrm{outdeg}(u)\ell)$ yields the following bound.

Theorem 8.12. *Consider a directed acyclic graph G. Let $\ell := \ell(G)$. Then the expected number of iterations of MMAS$_{SDSP}$ on G with $\tau_{\min}(u) = 1/(\mathrm{outdeg}(u) \cdot \ell)$ for all vertices $u \in V$ is $O(\ell|E| + (n \log n)/\rho)$.*

Proof. The structure of the proof is the same as that for Theorem 8.10. Only the expected time (number of iterations) until the vertices considered in topographical ordering have been processed is (are) added up more carefully. The expected time until vertex u has been optimized is at most $2e/\tau_{\min}(u) = 2e \, \mathrm{outdeg}(u)\ell$. Since $\tau_{\min}(u) \geq 1/n^2$, the processing time for vertex u is at most $2e \, \mathrm{outdeg}(u)\ell + \log(\tau_{\max}/\tau_{\min})/\rho = O(\mathrm{outdeg}(u)\ell + (\log n)/\rho)$. Adding this up over $n - 1$ vertices and noting that $\sum_{u \in V} \mathrm{outdeg}(u) = 2|E|$ yields the result.   □

8.3.2 All Pairs Shortest Paths

It is straighforward to compute shortest paths between all pairs of vertices by calling the MMAS$_{\mathrm{SDSP}}$ ACO algorithm sequentially for all n destination vertices. This can also be done in parallel by letting ants head for all destinations and extending the pheromone values on edges to vector-valued values. That is, we have ants $a_{u,v}$ and best-so-far paths $p_{u,v}^*$ for every start vertex u and every destination v. For every v, we introduce a pheromone function $\tau_v \colon E \to \mathbb{R}_0^+$

such that $\tau_v(e)$ denotes the pheromone value on edge $e \in E$ that controls the ants heading to v and is updated by these ants. The resulting algorithmic framework is called MMAS$_{\text{APSP}}$ and displayed in Algorithm 16. Note that the framework is nothing else than a parallelization of Algorithm 15. For this reason, it is easy to obtain results in the vein of Theorem 8.11 also for MMAS$_{\text{APSP}}$. Basically, the number of constructed solutions grows by a factor of n.

Algorithm 16 MMAS$_{\text{APSP}}$

For all $v \in V$, initialize pheromones τ_v and best-so-far paths $p^*_{1,v}, \ldots, p^*_{n,v}$.
loop
  for all $v \in V$ **do**
    for all $u \neq v$ **do**
      Construct a simple path $p_{u,v}$ from u to v with respect to τ_v.
      if $w(p_{u,v}) \leq w(p^*_{u,v})$ **then**
        $p^*_{u,v} \leftarrow p_{u,v}$
      end if
    end for
    Update pheromones with respect to $p^*_{1,v}, \ldots, p^*_{n,v}$.
  end for
end loop

Remarkable improvements over the plain parallelization can be obtained by introducing a small dose of interaction into MMAS$_{\text{APSP}}$. Consider an arbitrary ant $a_{u,v}$ heading for destination v. The idea is to give $a_{u,v}$ access to additional pheromone trails beside τ_v and let it follow these "foreign" trails with a certain probability. More precisely, the decision to follow a foreign trail is made with probability $1/2$. In this case, an intermediate destination $w \in V$ is chosen uniformly at random and the ant travels first to w using the pheromone information τ_w; afterwards it travels from w to the actual destination v using its own pheromone vector τ_v. This path construction with (possible) interaction is stated as Algorithm 17. The pheromone update for ant $a_{u,v}$ always applies exclusively to the pheromone values τ_v.

Theorem 8.13. *Consider a directed graph G with positive weights where all shortest paths are unique. Let $\ell := \ell(G)$ and $\Delta := \Delta(G)$. If $\ell \geq \ln n$ holds, the number of iterations of MMAS$_{APSP}$ with interaction on G with $\tau_{\min} = 1/(\Delta \ell)$ is $O(n \log n + \log(\ell) \log(\Delta \ell)/\rho)$ with probability at least $1 - 1/n^2$. The bound on the number of iterations holds also in expectation.*

Proof. As a preparation, we introduce concepts similar to those in the proof of Theorem 8.10. For an arbitrary pair of vertices (u, v), we denote by $\ell_{u,v}$ the maximum number of edges on a shortest path from u to v. We call an edge *incorrect* with respect to v if it does not belong to a shortest path to v. We call the pair (u, v) *optimized* if a shortest path from u to v has been found.

Algorithm 17 Path construction from u to v for $\text{MMAS}_{\text{APSP}}$ with interaction

Choose $b \in \{0, 1\}$ uniformly at random.
if $b = 0$ **then**
    Construct a simple path from u to v with respect to τ_v.
else
    Choose $w \in V$ uniformly at random.
    Construct a simple path $p' = (p'_0, \ldots, p'_{\ell'})$ from u to w with respect to τ_w.
    Construct a simple path $p'' = (p''_0, \ldots, p''_{\ell''})$ from w to v with respect to τ_v.
    if $p'_{\ell'} = w$ **then** $p \leftarrow (p'_0, \ldots, p'_{\ell'}, p''_1, \ldots, p''_{\ell''})$ **else** $p \leftarrow p'$ **end if**
end if
return p.

Finally, we call (u, v) *processed* if it has been optimized and the pheromone values $\tau_v(\cdot)$ on all incorrect edges $(u, \cdot)$, i.e., all incorrect edges leaving u, equal $\tau_{\min}$.

The actual proof divides the run of $\text{MMAS}_{\text{APSP}}$ into phases such that all pairs (u, v) with a certain bound on the $\ell_{u,v}$-value are processed in a phase. Since the bound increases by a factor $3/2$ from phase to phase, the total number of phases is bounded by $\alpha := \lceil \log(\ell)/\log(3/2) \rceil$. While going from one phase to the next, we exploit the ants' capability of using foreign pheromone trails and let them follow shortest paths with lower $\ell_{u,v}$-values between previously processed pairs. More precisely, the ith phase, $0 \le i \le \ell$, finishes when all pairs (u, v) satisfying $(3/2)^{i-1} < \ell_{u,v} \le (3/2)^i$ have been processed. Hence, the aim for the 0th phase is to process all pairs (u, v) such that $(u, v) \in E$.

We reserve $t^* := (\ln(2))/\rho$ iterations at the beginning of the first phase. We fix an arbitrary pair (u, v) such that $(u, v) \in E$. The probability of optimizing (u, v) in the tth iteration, $1 \le t \le t^*$, is at least $(1 - \rho)^{t-1}/(4\Delta)$ since the ant $a_{u,v}$ decides with probability $1/2$ to head directly for v (instead of following foreign pheromone trails) and chooses (u, v) with probability at least $(1 - \rho)^{t-1}/(2\Delta)$. The latter bound follows from Corollary 8.8 along with the fact that $\tau_v(u, v) \ge (1/\Delta) \cdot (1 - \rho)^{t-1}$ after $t - 1$ iterations. Hence, the probability of not optimizing (u, v) within the first t^* iterations is at most

$$\prod_{t=1}^{t^*} \left(1 - \frac{(1-\rho)^{t-1}}{4\Delta}\right) \le e^{-\frac{1}{4\Delta} \cdot \sum_{t=0}^{t^*-1}(1-\rho)^t} = e^{-\frac{1-(1-\rho)^{t^*}}{4\Delta\rho}}.$$

Since $\rho \le 1/(23\Delta \log n) \le 1/(8\Delta \ln(2n^4))$ and, therefore, $1 - (1-\rho)^t \ge 1/2$, the last probability is at most $e^{-\ln(2n^4)} = 1/(2n^4)$. By the union bound, the probability that there is a pair $(u, v) \in E$ left after the t^* iterations that is not optimized is at most $1/(2n^2)$. By Lemma 8.9 (which applies also to $\text{MMAS}_{\text{APSP}}$), all optimized pairs become processed after at most $\ln(\tau_{\max}/\tau_{\min}/\rho)$ iterations. Hence, the total length of the first phase is chosen to be $t^* + \ln(\tau_{\max}/\tau_{\min})/\rho$ iterations.

Consider phase i, $i \geq 1$, where all pairs (u', v') with $\ell_{u',v'} \leq (3/2)^{i-1}$ have already been processed. Now let (u, v) be a pair where $(3/2)^{i-1} < \ell_{u,v} \leq (3/2)^i$. If ant $a_{u,v}$ decides to use foreign pheromone information to head for an intermediate vertex w and chooses w on the middle third of a shortest path $p_{u,v}$ from u to v, then both $\ell_{u,w} \leq (3/2)^{i-1}$ and $\ell_{w,v} \leq (3/2)^{i-1}$ hold. The probability of choosing w in the desired way is at least $(1/2) \cdot (1/3) \cdot (\ell_{u,v}/n)$. In this case, the ant follows the shortest path from u to v via w with a probability of at least $(1-1/\ell)^{\ell-1} \geq 1/e$ since (u, w) and (w, v) have been processed by assumption. Altogether, the probability of optimizing (u, v) in a single iteration of phase i is at least $(1/2) \cdot (1/3) \cdot (\ell_{u,v}/n) \cdot (1/e) \geq (3/2)^{i-1}/(6en)$. By reserving $t_i^* := 6en \ln(2\alpha n^4)/(3/2)^{i-1}$ iterations for phase i, the probability of not optimizing a pair (u, v) with $(3/2)^{i-1} < \ell_{u,v} \leq (3/2)^i$ in the phase is at most $(1 - (3/2)^{i-1}/(6en))^t \leq 1/(2\alpha n^4)$. By the union bound, all such pairs are optimized within t_i^* iterations with probability at least $1 - 1/(2\alpha n^2)$. We already know that optimized pairs become processed after at most $\ln(\tau_{\max}/\tau_{\min})/\rho$ additional iterations. Altogether, the number of iterations reserved for all phases is at most

$$\frac{\ln 2}{\rho} + \sum_{i=1}^{\alpha} \left(\frac{6en \ln(2\alpha n^4)}{(3/2)^{i-1}} \right) + \alpha \cdot \frac{\ln(\tau_{\max}/\tau_{\min})}{\rho}$$

$$\leq \frac{\ln 2}{\rho} + 6en \ln(2\alpha n^4) \cdot \sum_{i=1}^{\alpha} \left(\frac{2}{3} \right)^{i-1} + \frac{\alpha \ln(\Delta \ell)}{\rho}$$

$$= O(n \log n + \log(\ell) \log(\Delta \ell)/\rho).$$

Summing over all phases, the total failure probability is at most $1/(2n^2) + \alpha \cdot 1/(2\alpha n^2) = 1/n^2$. This proves the first statement of the theorem. The second statement follows since the weaker upper bound $O(n^3 + (n \log n)/\rho)$ on the expected number of iterations of MMAS$_{\text{APSP}}$ follows easily using the proof ideas for Theorems 8.10 and 8.11. In the case where the time bounds set up for the phases are not sufficient, we estimate the number of iterations by the weaker bound. The contribution to the total expected number of iterations is at most $(1/n^2) \cdot O(n^3 + (n \log n)/\rho) = O(n \log n + \log(\ell) \log(\Delta \ell)/\rho)$.   □

Since each iteration of MMAS$_{\text{APSP}}$ constructs n^2 solutions, the bound of the last theorem corresponds to $O(n^3 \log n + n^2 \log(\ell) \log(\Delta \ell)/\rho)$ evaluations of the objective function, which is an optimization time of $O(n^3 \log n)$ if ρ is chosen appropriately. This beats the bound for Steady State GA$_{\text{SP}}$ presented in Theorem 8.6 (and also the lower bound $\Omega(n^{3.25} \log^{1/4} n)$ given by Doerr and Theile (2009)). Hence, the ACO variant can be called the most efficient algorithm among the stochastic search algorithms considered in this chapter.

Conclusions

We have considered different search algorithms for the computation of shortest paths. For the single-source shortest-path (SSSP) problem, a simple evolutionary algorithm was investigated on two different fitness functions. With respect to the all-pairs shortest-paths (APSP) problem, a population-based evolutionary algorithm using crossover is compared to the same algorithm without crossover and a benefit of the crossover-based variant is proven. Finally, both the SSSP and the APSP were solved by ACO algorithms with a nontrivial colony. A variant using interaction leads to even better runtime bounds than the ones presented for evolutionary algorithms.

Eulerian Cycles

In this chapter, we analyze stochastic search algorithms on arc routing problems. For such problems, the choice of a good representation is not straightforward, and it has a large impact on the success of stochastic search algorithms. The Eulerian cycle problems is the simplest problem belonging to the wide class of arc routing problems, and we consider this problem as an example of how the choice of the representation influences the runtime of stochastic search algorithms.

Euler initiated the study of graph theory with the famous seven bridges problem (Euler, 1741). The generalization of the seven bridges problem can be described as follows and is known as the Eulerian cycle problem. Given an undirected connected graph $G = (V, E)$ on n vertices and m edges, the task is to compute a cycle such that every edge is used exactly one time. Euler proved that a tour of all edges in a connected undirected graph without repetition is possible iff the degree of each vertex is even. Such graphs are known as Eulerian graphs. If an Eulerian cycle exists, we call G Eulerian. In the rest of this chapter, we assume that G is Eulerian.

The Eulerian cycle problem can be solved in time $O(m + n)$ by the algorithm of Hierholzer (1873) (see Algorithm 18). This algorithm computes cycles in the given graph and joins them together such that an Eulerian tour is obtained.

Stochastic search algorithms do not have the knowledge that the problem can be solved by computing cycles and building up the solution by putting the cycles together. We will see that they are able to compute a single cycle and integrate another cycle if the solution is not optimal. Hence, they follow the idea of the algorithm without having this global knowledge.

We do not and cannot hope to compete with the best algorithms for the Eulerian cycle problem. This can be different for generalizations of the problem. For example, the problem of finding the largest Eulerian subgraph of a given graph and the mixed Chinese postman problem (see Edmonds and Johnson, 1973) are NP-hard and stochastic search algorithms have a good chance to be competitive on these problems. For other NP-hard variants such

F. Neumann, C. Witt, *Bioinspired Computation*
in Combinatorial Optimization, Natural Computing Series,
DOI 10.1007/978-3-642-16544-3_9, © Springer-Verlag Berlin Heidelberg 2010

Algorithm 18 Algorithm of Hierholzer

---

1. Find a cycle C in G
2. Delete the edges of C from G
3. If G is not empty go to step 1.
4. Construct the Eulerian cycle from the cycles produced in Step 1.

---

as the capacitated arc routing problem, evolutionary algorithms have been developed and successfully applied (Lacomme, Prins, and Ramdane-Chérif, 2001).

We consider stochastic search algorithms that use different representations to find an Eulerian cycle. The representation of permutations has successfully been applied to difficult combinatorial optimization problems such as the traveling salesperson problem (see Michalewicz and Fogel, 2004 for an overview). We start with this general approach as it is important to understand how evolutionary algorithms, using this encoding, work on simple problems. Later on, we investigate a representation based on adjacency list matchings, which is more related to the Eulerian cycle problem, and show that it leads provably to a better runtime behavior of stochastic search algorithms.

9.1 Edge Permutations

In this section, we examine how the general approach of representing solutions by permutations of the edges can solve the Eulerian cycle problem.

9.1.1 Algorithms

To find such a cycle, we use a permutation of the edges of G. The search space S_m contains all permutations of the edges of G. A search point $\pi \in S_m$ corresponds to the order of using the edges for the Eulerian tour. Usually a permutation does not correspond to an Eulerian tour. It normally describes a walk w which is part of such a tour. The ideas can be used to define the fitness function *walk*, which is appropriate for the Eulerian cycle problem.

The fitness of a permutation π is given by

$$walk(\pi) := \text{length of the walk implied by } \pi,$$

where we start with the first edge in π and extend the walk if the edge on the second position has one vertex with the first edge of π in common. This walk can be further extended if the third edge has one vertex which is equal to the "free" vertex of the second edge. We can extend the procedure to build up a walk of length ℓ implied by π. In the rest of this section, the walk will be named by w. Usually a walk w is written as a sequence of vertices and

---

Algorithm 19 RLS$_p$

1. Choose $\pi \in S_m$ randomly.
2. Choose i und j uniform at random and define π' by executing $jump(i, j)$ on π.
3. Replace π with π' if $walk(\pi') \geq walk(\pi)$.
4. Repeat Steps 2 and 3 forever.

Algorithm 20 RLS$_a$

1. Choose $\pi \in S_m$ randomly.
2. Choose i uniform at random and define π' by executing $jump(i, 1)$ on π.
3. Replace π with π' if $walk(\pi') \geq walk(\pi)$.
4. Repeat Steps 2 and 3 forever.

---

denoted by $w = (v_0, v_1, \ldots, v_\ell)$. This implies a set of edges that is a subset of the edge set E. To make the connection to the fitness function $walk$ more precise, we represent a walk w by a sequence of directed edges and denote it by $w = (v_0, v_1), (v_1, v_2), \ldots, (v_{\ell-1}, v_\ell)$.

The fitness function describes the processing order in which to use the edges for a tour starting with the edge on position 1. The fitness of a permutation can therefore be easily evaluated. If the resulting walk is short, most edges in the permutation do not have to be considered.

In the case where we are searching for a good permutation of the input elements, jumps and exchanges are popular operators that lead to new solutions (see Section 3.1.2). Both operators have been integrated into one mutation operator by Scharnow et al. (2004) for the sorting problem. We consider the jump operator in this section and show that it leads to an efficient optimization process.

We investigate a variant of RLS with permutations of the edges. The algorithm RLS$_p$ executes in one mutation step exactly one jump operation. This jump is chosen according to the uniform distribution from among all possible jumps, which means that the positions i and j are chosen uniformly at random from the set $\{1, \ldots, m\}$. Our algorithm starts with a permutation π chosen randomly from the set S_m that consists of all permutations of m elements. We will analyze such stochastic search algorithms until they have found a good permutation of the edges of a given graph for the Eulerian cycle problem. The underlying fitness function $walk$ should be maximized. Therefore, we describe RLS$_p$ as shown in Algorithm 19.

We also want to examine whether a slight modification of the mutation operator leads to significantly better results. The idea is to use asymmetric jumps which choose one edge uniformly at random and place this edge at the first position of the permutation. The algorithm RLS$_a$ (Algorithm 20) differs from RLS$_p$ by using this asymmetric mutation operator.

Working with a mutation operator that allows more than one jump operation in a single mutation step, we obtain variants of (1+1) EA for the Eulerian cycle problem. All results given in this section can be generalized to a variant of (1+1) EA, where the number of jump operations in a single mutation operation is chosen according to a Poisson distribution with parameter $\lambda = 1$. We do not want to give the full proofs (which can be found in Neumann, 2008 and Doerr, Hebbinghaus, and Neumann, 2007b) as they follow similar ideas as those for RLS_p and RLS_a, but involve more technicalities.

9.1.2 Runtime Analysis

In the following, we show an upper bound of $O(m^5)$ on the expected optimization time for RLS_p on the proposed fitness function.

Theorem 9.1. *The expected time until RLS_p working on the fitness function walk constructs an Eulerian cycle is $O(m^5)$.*

Proof. The fitness $walk(\pi)$ of a search point π can take values from $\{1, \ldots, m\}$, where the optimum is reached if $walk(\pi)$ equals m. Our goal is to show that an improvement, i.e., a solution of fitness at least $\ell + 1$, has been obtained after an expected number of $O(m^4)$ steps.

W. l. o. g., the walk w implied by π is of the form $(v_0, v_1), \ldots, (v_{\ell-1}, v_\ell)$ and has length $\ell \in \{1, \ldots m - 1\}$. If $v_0 \neq v_\ell$ holds, we say that w is a path, and otherwise we say that w is a cycle. In the first case, we consider a typical run consisting of a phase of cm^2 steps, c an appropriate constant, and show that an improvement has been reached with probability $\Omega(1)$ within this phase. Hence, the expected number of such phases to reach an improvement is upper bounded by α^{-1}. In the second case, we show that either an improvement has been reached in expected time $O(m^4)$ or a path of length ℓ has been produced. If a path of length ℓ has been produced before the improvement, we may return to the first case.

Claim. Let w be a path described by π of length $\ell \in \{1, \ldots, m - 1\}$. Considering a phase of length cm^2, c an appropriate constant, an improvement has been achieved with probability at least $\alpha = \Omega(1)$.

Proof. We work under the assumption that w has not been turned into a cycle of length ℓ before having achieved an improvement. If $v_0 \neq v_\ell$ holds, there is an edge incident to v_ℓ which can be placed after $\{v_{\ell-1}, v_\ell\}$. Such a jump lengthens the path and has probability at least $1/m^2$. In this case, the expected time for an improvement of π is bounded by m^2. Using Markov's inequality, the probability that that an improvement has been achieved within a phase of cm^2 steps is at least $1 - 1/c$.

However, it might happen that the path turns into a cycle of length ℓ before an improvement has been obtained. This can happen if the edge $e = \{v_{\ell-1}, v_0\}$

is put on position ℓ (assuming that it exists in the set of edges E). Our goal is to bound the probability for this event in the following.

Let k be the position of e in π. The operation $jump(k, \ell)$ takes e and puts it directly in position ℓ. The probability for this event is $1/m^2$. Therefore, the probability of an improvement is at least as high as the the probability of executing this jump. Hence, an improvement happens before this jump with probability at least $1/2$.

If $k = \ell + 1$, $jump(\ell, *_1)$, where $*_1 \in \{\ell + 1, \ldots, m\}$, can be executed such that e is put in position ℓ. The probability of this operation is $\frac{m-\ell}{m}$. However, with probability $\frac{m-\ell-1}{m}$, $jump(k, *_2)$ with $*_2 \in \{\ell + 2, \ldots, m\}$ is executed. After this has happened, the probability of putting e in position ℓ is again $1/m^2$, and an improvement happens with a probability of $1/2$ before this event. The probability of executing $jump(k, *_2)$ before $jump(\ell, *_1)$ is $\frac{m-\ell-2}{2m-2\ell-3} \geq 1/3$.

Altogether with probability at least $\alpha = (1 - 1/c) \cdot \frac{1}{6} = \Omega(1)$, an improvement has been achieved during the considered phase. □

If $v_0 = v_\ell$ the analysis for an improvement is more complicated. In this case, w is a cycle C. If the graph is Eulerian and w is not an Eulerian tour, there is at least one vertex v_k on C which is also a vertex on another cycle C' having v_k in common with C (see Figure 9.1). We want to show that an improvement is reached in expected time $O(m^4)$.

Claim. Let w be a cycle described by π of length $\ell \in \{1, \ldots, m - 1\}$. The expected time to produce an improvement is $O(m^4)$.

Proof. We bound the time the reach an improvement by $O(m^4)$ when starting with a cycle under the condition that no path of length ℓ has been produced before. If we reach a path of length ℓ before the improvement, we already know that after an additional phase of length cm^2, an improvement happens with probability $\alpha = \Omega(1)$. This implies that an improvement is reached in expected time $\alpha^{-1}(O(m^4) + O(m^2)) = O(m^4)$ when starting with a cycle.

We inspect the case where w is a cycle whose corresponding permutation π does not start with the vertex v_k. We call an operation relevant if it changes w and is accepted.

1. Case $i, j \in \{\ell + 2, \ldots, m\}$: These operations are not relevant as they do not change w.
2. $i \in \{1, \ldots, \ell+1\}, j \in \{\ell+2, \ldots, m\}$: If $i \in \{1, \ldots, \ell\}$, the cycle is destroyed. If $i = \ell$ and the edge at position $\ell + 1$ contains $v_{\ell-1}$, a path of length ℓ is constructed. If $i = \ell + 1$, an improvement may happen if the edge at position $\ell + 2$ contains v_ℓ. Otherwise, w is unchanged.
3. $i \in \{\ell + 2, \ldots, m\}$ and $j \in \{1, \ldots, \ell + 1\}$: If $j \in \{2, \ldots, \ell\}$, the cycle property is destroyed and either the fitness is decreased or a path is constructed. If $j = \ell + 1$, an improvement is reached if the edge at position i contains v_ℓ. The same holds for $j = 1$ if the edge at position i contains v_0. The walk w remains unchanged for all other cases where $j = \ell + 1$ holds.

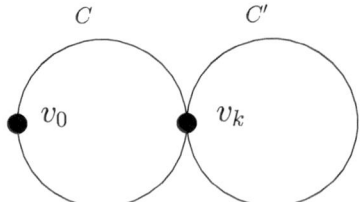

Fig. 9.1. Situation in which w is a cycle C that does not include all edges of G. Then there is another cycle C' which has one vertex v_k with C in common

4. $i, j \in \{1, \ldots, \ell + 1\}$: If i or $j \in \{2, \ldots, \ell - 1\}$, we destroy the cycle. These steps are not accepted by the algorithm.

    If $i = \ell + 1$ and $j = 1$, an improvement may happen if the edge on position $\ell + 1$ contains v_0. If $i = \ell + 1$ and $j = \ell$, the cycle may be turned into a path of length ℓ.

    An operation where $i = 1$ and $j \neq \ell$ or $i = \ell$ and $j \neq 1$ shortens the walk by at least 1 and is not accepted. The algorithm accepts the two jump operations $jump(1, \ell)$ and $jump(\ell, 1)$. These operations revolve the cycle.

The only two relevant operations that produce from a cycle w another cycle w' are $jump(1, \ell)$ and $jump(\ell, 1)$. All other relevant operations produce a path of length at least ℓ.

If w is a cycle C which is not an optimal solution, then there is at least another cycle C' that shares with C at least one vertex v_k.

To reach an improvement we examine how to construct a cycle

$$w^* = (v_k, v_{k+1}), \ldots, (v_{\ell-1}, v_0), \ldots, (v_{k-1}, v_k).$$

We investigate how this can be done by the two jump operations $jump(1, \ell)$ and $jump(\ell, 1)$. If we have not reached such a walk, there is exactly one jump which places the edge $e = \{v_k, v_{k+1}\}$ one position further to the left and one which places e one position further to the right. The probability of placing e further to the left is in each relevant step $\frac{1}{2}$. Hence, the algorithm performs a random walk shifting the edge e to the left or to the right with equal probability. Using the results on fair random walks presented in Section 4.2.4, the expected number of relevant steps when starting with e in the permutation is $O(m^2)$ as the number of edges in w is upper bounded by m. Each relevant step happens with probability $1/m^2$, which implies that an expected number of $O(m^4)$ steps suffices when starting with e.

The vertex v_k is also a vertex in another cycle C'. Hence, there are two edges $\{v_k, v_s\}$ and $\{v_k, v_t\}$ in C'. If we place one of these edges at position $\ell + 1$ or 1 (see 3.), we have lengthened the walk and achieved an improvement. On the other hand, the operations $jump(1, \ell)$ and $jump(\ell, 1)$ put e at position ℓ and position 2, respectively. This yields a probability of at least $\frac{1}{2}$ for an improvement in the next relevant step. Therefore, the algorithm has to produce a cycle starting with v_k at most twice in expectation, which implies that the expected time to reach an improvement is $O(m^4)$. □

There are at most $m - 1$ improvements, which leads to an upper bound of $O(m^5)$ on the expected optimization time of $\mathrm{RLS_p}$. □

In Doerr et al. (2007b), it has been shown that $\mathrm{RLS_p}$ needs an expected number of $\Omega(m^4)$ steps to find an Eulerian cycle if the given graph consists of two cycles that contain $m/2$ edges and share one single vertex v. The reason for this lower bound is that the random walk revolving the cycle needs $\Omega(m^4)$ to produce a permutation that starts with an edge containing v.

We want to discuss whether using asymmetric jumps leads to more efficient evolutionary algorithms and examine $\mathrm{RLS_a}$ in the following. Using an asymmetric jump operation has the following effect. In the case where the current solution is a path, the upper bound on the expected waiting time to lengthen the walk reduces from $O(m^2)$ to $O(m)$. In the case where the current solution is a cycle, the algorithm performs a directed walk instead of a random walk. This reduces the expected time for an improvement significantly, as shown in the following theorem.

Theorem 9.2. *The expected time until $\mathrm{RLS_a}$ has computed a Eulerian cycle is bounded by $O(m^3)$.*

Proof. $\mathrm{RLS_a}$ executes only jumps to the first position in the permutation. We assume that our current solution represents a walk of length ℓ. Again, we assume that the walk is of the form $(v_0, v_1), \ldots, (v_{\ell-1}, v_\ell)$. If $\ell \geq 2$ holds, a jump is only accepted if the edge $e = (v_i, v_j)$ at position i, which is jumped to position 1, contains v_0 or v_1. If the edge e contains v_1, a path of length at most 2 is obtained as $v_0 \cap e = \emptyset$. If e contains v_0, this may either lead to an improvement if $e \neq \{v_{\ell-1}, v_\ell\}$ or revolve the cycle.

We distinguish between two cases. In the first case, the current walk is a path. Then there exists at least one edge $e \neq \{v_{\ell-1}, v_\ell\}$ that can jump to position 1 and lengthen the walk. In this case, the expected waiting time for an improvement is $O(m)$.

In the second case, w is a cycle C that is not a Eulerian cycle. Then there is at least another cycle C' that shares a vertex v_k with C. As for $\mathrm{RLS_p}$, we consider the time to construct a walk

$$w^* = (v_k, v_{k+1}), \ldots, (v_{\ell-1}, v_0), \ldots, (v_{k-1}, v_k).$$

This is achieved by executing $\ell - k$ times the operation $jump(\ell) := jump(\ell, 1)$. Each of these jumps happens with probability $1/m$, and the expected time to produce a cycle which starts with v_k is therefore at most $(\ell - k)m = O(m^2)$. Afterwards, an improvement can be achieved by jumping one of the edges $\{v_k, v_s\}$ and $\{v_k, v_t\}$ contained in C' to position 1. On the other hand, $jump(\ell)$ is also accepted and revolves the cycle further. However, the probability that the next accepted mutation step is an improvement is at least 2/3. Altogether, the expected time for an improvement is upper bounded by $O(m^2)$ in the second case. The number of improvements is at most $m-1$, which completes the proof. □

9.2 Adjacency List Matchings

In the previous section, we have shown that simple stochastic search algorithms representing possible solutions as permutations of the edges achieve an Euler tour of a given Eulerian graph in expected polynomial time. In the following, we want to examine how representations that are more related to the given problem speed up the optimization process.

The fitness function *walk* considers a walk starting with the first edge in the permutation. This walk is extended as long as possible, resulting in the fitness value of a permutation of the edges. The idea that leads to a more efficient optimization process is to consider not only one specific walk, but a set of walks which can be merged such that an Eulerian tour is obtained. This idea has been used by Doerr and Johannsen (2007), who have chosen a representation of possible solutions for the Eulerian cycle problem based on adjacency list matchings. We will see that this representation in combination with suitable mutation operators leads to significantly improved runtime bounds.

9.2.1 Algorithms

The representation by adjacency list matchings is based on a phenotype-genotype mapping which is often used in evolutionary algorithms. Solutions are represented in the genotype space and are mapped via a specific function to the phenotype. The Eulerian cycle problem consists of finding a tour in the graph such that each edge is used exactly once. Therefore, it seems appropriate that solutions in the phenotype space consist of different edge-disjoint paths and cycles in the graph.

Phenotype

We first describe how these ideas are represented in the phenotype and discuss the genotype representation afterwards. A phenotype consists of different walks that can later be joined such that an Eulerian tour is obtained. A path of length k is given by a sequence of vertices $(v_0, v_1, \ldots, v_k)$ such that there is an edge $e_i = \{v_{i-1}, v_i\} \in E$, $1 \le i \le k$, and $v_0 \ne v_k$ holds. Similarly, a cycle of length $k+1$ consists of a sequence of vertices $(v_0, v_1, \ldots, v_k, v_0)$ such that there is an edge $e_i = \{v_{i-1}, v_i\}$, $1 \le i \le k$ and $e_{k+1} = \{v_k, v_0\}$. As we consider the Eulerian cycle problem, we require that all edges be different, but allow that vertices appear more than once.

A phenotype is a cover $C = \{C_1, \ldots, C_r\}$ for the given graph. It consists of a set of r edge-disjoint paths and cycles which cover all edges of G. The

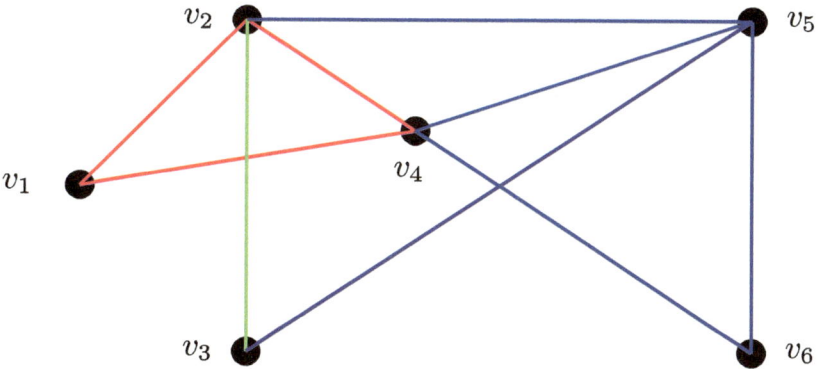

Fig. 9.2. Graph G with six vertices and nine edges and four edge-disjoint walks,

fitness of a cover $C = \{C_1, \ldots, C_r\}$ is given by the number of edge-disjoint walks, i.e.,

$$f(C) = r.$$

An optimal solution has fitness 1 as it uses all edges in a single walk, which implies that this walk is an Eulerian cycle.

Genotype

We have to give a suitable representation for the genome. This is based on adjacency lists, which are a common representation for graph problems. For each vertex $v \in V$, we store its neighbors in G in a list A_v. As the graph is Eulerian, the number of entries in each list is even. In total, we have n lists and the total number of entries in all lists is $2m$ as each edge $\{u, v\}$ contributes an entry u to list A_v and an entry v to list A_u.

The idea behind the adjacency list matching is that we can match two vertices u and w in a list A_v such that a path u, v, w of length 2 is obtained. A matching M_v of a list A_v is a set of disjoint pairs of vertices. M_v is called *perfect* iff all vertices in A_v are matched. A matching $M = \cup_{v \in V} M_v$ is a set of n matchings, one for each list A_v. M is called perfect if for each $v \in V$, M_v is a perfect matching.

In Doerr and Johannsen (2007), it is shown that there exists a mapping between the phenotype space of paths of G and the genotype space of matchings. This mapping is used to match search points in the genotype to a collection of paths in the phenome. We present the theorem giving the mapping from the genotype to the phenome in the following.

Theorem 9.3. *There exists a 1↔1 correspondence between the phenotype space and the genotype space. Moreover, this 1↔1 correspondence maps cycle covers of G to perfect matchings and vice versa.*

Proof. The idea behind the mapping is that unmatched vertices correspond to end vertices of a path and matched vertices to interior vertices of paths and cycles.

First, we show that a walk cover defines a matching M. Let $C = \{C_1, \ldots, C_r\}$ be a cover consisting of r walks (either paths or cycles), where each edge $e \in E$ only occurs at most once in all walks. For a vertex $v \in V$ and $u, w \in A_v$, $\{u, w\} \in M_v$ holds iff there exists a walk or a cycle C_i such that the $\{u, v\}$ and $\{v, w\}$ are subsequent in C_i. Each edge $e = \{u, v\}$ appears exactly once in C (u once in A_v and v once in A_u). Hence, M is a matching.

Now, we show that a matching M defines a walk cover C of G. The empty matching defines of a set of m paths u, v corresponding to the edges $\{u, v\} \in E$ in G. Let M be a non-empty matching, $v \in V$ and $\{u, w\} \in M_v$. We define recursively a walk cover C corresponding to M. Let $M'_v = M_v \setminus \{u, w\}$ and $M' = (M \setminus M_v) \cup M'_v$ and let C' be the cover corresponding to M'. Since u and w are not matched in M', there exist either two walks $(v', \ldots, u, v)$ and $(v, w, \ldots, v'')$ or one path $(v, u, \ldots w)$ in C'. In the first case, the walk cover C is defined as the walk cover C', where the two walks are joined to one walk $(v', \ldots, u, v, w, \ldots, v'')$. In the second case, the walk cover C is defined as C', where the path is replaced by the cycle $(v, u, \ldots, w, v)$. $\square$

Due to this 1↔1 correspondence, we identify a matching M with its walk cover C and vice versa.

We want to illustrate the use of adjacency list matchings and consider the Eulerian graph G given in Figure 9.2. Consider the matching $M = \cup_{i=1}^{6} M_{v_i}$ with

$$M_{v_1} = \{\{v_2, v_4\}\},$$
$$M_{v_2} = \{\{v_1, v_4\}\},$$
$$M_{v_3} = \emptyset,$$
$$M_{v_4} = \{\{v_1, v_2\}, \{v_5, v_6\}\},$$
$$M_{v_5} = \{\{v_2, v_6\}\},$$
$$M_{v_6} = \{\{v_4, v_5\}\}.$$

This matching corresponds to the set of walks

$$C = \{ \ (v_1, v_2, v_4, v_1),$$
$$(v_2, v_5, v_6, v_4, v_5),$$
$$(v_2, v_3),$$
$$(v_3, v_5)\}.$$

Mutation

The mutation operator works on the genome, i.e., it matches or unmatches vertices in the adjacency lists. It carries out a sequence of local operations. In

Algorithm 21 Local operation

Input: u, w, A_v

1. If u and w are unmatched, $M_v := M_v \cup \{u, w\}$.
2. If $\{u, w\} \in M_v$, $M_v := M_v \setminus \{u, w\}$.
3. If u is matched to some w' and w is unmatched,
   $M_v = (M_v \setminus \{u, w'\}) \cup \{u, w\}$.
4. If w is matched to some v' and u is unmatched,
   $M_v = (M_v \setminus \{w, v'\}) \cup \{u, w\}$.
5. If u is matched to some w' and w is matched to some u',
   $M_v = (M_v \setminus (\{u, w'\} \cup \{u', w\})) \cup (\{u, w\} \cup \{u', w'\})$.

Algorithm 22 (1+1) EA$_M$

1. Choose a matching M.
2. Define M' in the following way. Choose ℓ from a Poisson distribution with parameter $\lambda = 1$ and perform sequentially $\ell + 1$ randomly chosen edge-based mutation operations to produce M' from M.
3. Replace M with M' if $f(C') \leq f(C)$.
4. Repeat Steps 2 and 3 forever.

---

a local operation, two vertices u and w from an adjacency list A_v are chosen. A question that arises is about how to pick the vertices u and w that are used in the operation. We discuss the edge-based approach where u is chosen uniformly from all $2m$ vertices in all lists. Other approaches can be found in Doerr and Johannsen (2007).

Suppose that u is in list A_v. Then w is chosen uniformly at random from A_v. Hence, the probability of choosing a specific pair (u, w) in list A_v is $\frac{1}{2d(v)m}$, where $d(v)$ denotes the degree of vertex v in G. Having chosen u and w, a step of the local operation works as follows. If $u = w$, nothing is changed. Otherwise, an operation according to the different cases given in Algorithm 21 is executed. The important cases are 1 and 5, which are necessary for achieving the improvements by the stochastic search algorithm. Case 1 joins two walks into a single one. Case 5 joins two cycles into a single cycle.

Analyzing stochastic search algorithms based on adjacency list matchings in Section 9.2.2, we will also consider the special case where we always work with perfect matchings. In this case, nothing is changed if $\{u, w\} \in M_v$ as deleting $\{u, w\}$ from M_v would destroy the property of having a perfect matching.

Finally, we can describe a variant of the (1+1) EA (see Algorithm 22) that works in the genotype space. Possible solutions are represented as adjacency list matchings. The mutation operator carries out sequentially a number of $\ell + 1$ local operations, where ℓ is chosen from a Poisson distribution with parameter $\lambda = 1$ to produce an offspring M'.

9.2.2 Runtime Analysis

We first consider (1+1) EA_M which always works with a perfect matching. A perfect matching represents different cycles in the phenoype that have to be joined during the optimization process. We are not allowed to produce solutions consisting of unmatched vertices in this case. Hence, only local operations due to Case 5 in Algorithm 21 are executed. As there is a $1 \leftrightarrow 1$ correspondence, between matchings and walk covers we identify a matching M with its walk cover C.

Theorem 9.4. *The expected optimization time of (1+1) EA_M working with perfect matchings is $O(m \log m)$.*

Proof. The algorithm works at each time step with a perfect matching. All walks are cycles. Hence, only the number of different cycles determines the fitness. Assume that the number of cycles of the current solution M is r. We want to compute a lower bound on the probability of obtaining a matching of at most $r - 1$ cycles in the next step and use the method of fitness-based partitions (see Section 4.2.1) according to the different possible values of r afterwards.

To reduce the number of cycles, the mutation operator has to join two cycles into one. We know that each cycle shares at least one vertex v with another cycle as the graph is Eulerian. Consider such a vertex v and assume that $s(v)$ cycles $c_1, \ldots, c_{s(v)}$, $s \geq 2$, have this vertex in common. We count the number of pairs (u, w) in A_v such that the edges $\{u, v\}$ and $\{v, w\}$ are in different cycles of $c_1, \ldots, c_{s(v)}$. Let d_i be the number of vertices in c_i, $1 \leq i \leq s(v)$, that are incident to vertex v. We know that $d_i \geq 2$ and $\sum_{i=1}^{s(v)} d_i = d(v)$. The total number of pairs (u, w) for list A_v is $d(v)^2$, and the number of pairs where both vertices belong to cycle c_i is d_i^2. Hence, the number of pairs that are in different cycles is

$$d(v)^2 - \sum_{i=1}^{s(v)} d_i^2 \geq s(v) \cdot d(v).$$

Therefore, the probability that two cycles sharing the vertex v are joined is at least

$$s(v) \cdot d(v) \cdot \frac{1}{2d(v)m} = \frac{s(v)}{2m}.$$

Let $V' \subseteq V$ be the set of vertices that are contained in at least two of the cycles given by M. Clearly, $\sum_{v \in V'} s(v) \geq r$. This implies that the probability of joining two of the r cycles in the next step is at least $r/(2m)$. A perfect matching consists of at most $m/3$ cycles. Using the method of fitness-based partitions, the expected optimization time is upper bounded by

$$\sum_{r=2}^{m/3} 2em/r = O(m \log m). \quad \square$$

When working with perfect matchings during the whole optimization process, the evolutionary algorithm has to join the different cycles and produces an Eulerian cycle in expected time $O(m \log m)$. This shows that joining cycles is much easier for the representation by adjacency list matchings than by the standard approach, which uses a permutation of the edges.

The general approach for an evolutionary algorithm using adjacency list matchings would be to start with an arbitrary matching and evolve this into a perfect matching describing an Eulerian cycle. We want to examine the runtime of this approach in the following.

Theorem 9.5. *The expected optimization time of (1+1) EA_M working with arbitrary matchings is $O(\Delta(G)m \log m)$, where $\Delta(G)$ denotes the maximum degree of the given input graph G.*

Proof. Let M be a matching whose corresponding cover C has fitness $f(C) = r, 1 \leq r \leq m$. In the following, we show that there are $\lfloor r/2 \rfloor$ operations, where each one improves the fitness of C by at least 1 and happens with probability $\Omega(\frac{k}{m\Delta(G)})$. Afterwards, we use the method of fitness-based partitions to obtain the upper bound on the expected optimization time.

We distinguish between paths and cycles of the matching M and examine how to improve the fitness of the current solution. First, we consider the case of a path. Let u be an end vertex of such a path and A_v the list in which u is not matched. The number of vertices in A_v is even, which implies that there is another unmatched vertex w in A_v. Setting $M_v = M_v \cup \{u, w\}$ reduces the number of walks by 1. Consider the case of a cycle defined by M. As G is Eulerian, there exists a least one vertex v in the cycle which is also contained in another path or cycle. Hence, there exists a mutation that connects the cycle to another component and reduces the fitness by at least 1.

As each of the r walks has at least one local operation that joins two components, the number of different mutations leading to an improvement is at least $\lfloor r/2 \rfloor$. Hence, $f(C) = r$ implies that the probability of an improvement in the next step is $\Omega(\frac{r}{\Delta(G)m})$. Using the method of fitness-based partitions according to the different values of r, the expected optimization time is upper bounded by $O(m\Delta(G) \log m)$. $\square$

Conclusions

The Eulerian cycle problem is a fundamental problem in graph theory belonging to the class of arc routing problems. Several important problems belonging to this class are difficult and stochastic search algorithms have a good chance of being competitive on these problems. We examined how such algorithms working with different representations of possible solutions can deal with the basic Eulerian cycle problem. Our results showed that a general approach representing possible solutions by a permutation of the edges of the input graph leads to an expected optimization time when using jump operations in the mutation operator. Later on, we examined a more problem-specific representation based on adjacency list matchings and showed that this leads provably to more efficient stochastic search algorithms.

Part III

Multi-objective Optimization

10

Multi-objective Minimum Spanning Trees

In this chapter, we analyze multi-objective evolutionary algorithms (MOEAs) on an NP-hard multi-objective combinatorial optimization problem, namely the multi-objective minimum spanning tree problem. Many successful evolutionary algorithms have been proposed for this problem (Knowles and Corne, 2001; Zhou and Gen, 1999). In Chapter 5, we showed that stochastic search algorithms are able to compute minimum spanning trees in expected polynomial time. The analysis is based on the investigation of the expected multiplicative distance decrease (where the distance is measured as the weight difference between the current solution and an optimal one) and serves as a starting point for the analysis of the multi-objective minimum spanning tree problem.

The problem of computing multi-objective minimum spanning trees can be stated as follows. Given an undirected connected graph $G = (V, E)$ on n vertices and m edges and for each edge $e \in E$ a weight vector $w(e) = (w_1(e), \ldots, w_k(e))$, where $w_i(e)$, $1 \le i \le k$, is a positive integer, the goal is to find for each objective vector q of the Pareto front F a spanning tree s with $w(s) = q$. In the case of at least two weight functions, the problem is NP-hard (see Ehrgott, 2005). Papadimitriou and Yannakakis (2000) have given a fully polynomial-time approximation scheme (FPTAS) to compute an ϵ-approximation of the Pareto front. This algorithm is based on a pseudo-polynomial algorithm given by Barahona and Pulleyblank (1987). The results given in these papers make use of matrix multiplication algorithms and do not give insights into how the problem may be tackled by using stochastic search algorithms. In this chapter, we consider the case $k = 2$ and examine which parts of the Pareto front can be computed by simple MOEAs in pseudo-polynomial time. The results we present are due to Neumann (2007).

The outline of this chapter is as follows. In Section 10.1, we introduce the algorithms that are subject to our investigations. In Section 10.2, we show that the extremal points give a 2-approximation of the Pareto front. We analyze GSEMO in Section 10.3 with respect to the expected time until it has produced a population that includes for each extremal point of the Pareto front a corresponding spanning tree, and finish with some conclusions.

F. Neumann, C. Witt, *Bioinspired Computation*
in Combinatorial Optimization, Natural Computing Series,
DOI 10.1007/978-3-642-16544-3_10, © Springer-Verlag Berlin Heidelberg 2010

10.1 Representation

Again, we use the edge set encoding for our algorithms. The search space equals $S = \{0,1\}^m$, where each position corresponds to one edge. A search point s corresponds to the choice of all edges e_j, $1 \le j \le m$, where $s_j = 1$ holds. Let $w_i^{\max}$ be the maximum weight of w_i, $w_{\max} = \max w_i^{\max}$, $w_{\min} = \min w_i^{\max}$, and $w_{\text{ub}} = n^2 \cdot w_{\max}$. The fitness of an individual s is described by a vector $f(s) = (f_1(s), \ldots f_k(s))$ with

$$f_i(s) := ((c(s), e(s) - (n-1), w_i(s)),$$

where $w_i(s) := \sum_{j=1}^m w_i^j s_j$ and w_i^j is the value of edge e_j with respect to the function w_i, $c(s)$ is the number of connected components in the graph described by s, and $e(s)$ is the number of edges in this graph. Each f_i should be minimized with respect to the lexicographic order. Note that the number of connected components and the number of chosen edges is the same for a particular search point s in each objective function f_i.

The fitness function f is a generalization of the first fitness function used for $\text{RLS}_b^{1,2}$ and $(1+1)$ EA_b in Chapter 5 to the multi-objective case. Again, the most important issue is to decrease $c(s)$ until we have graphs connecting all vertices. The next issue is to decrease $e(s)$ under the condition that s describes a connected graph. Finally, we look for Pareto optimal spanning trees.

The fitness function f penalizes the number of connected components as well as the extra connections. This is not necessary since breaking a cycle decreases the fitness value. Therefore, we are also interested in the fitness function $f'(s) = (f'_1(s), \ldots f'_k(s))$ with

$$f'_i(s) := ((c(s), w_i(s)),$$

which generalizes the second function of Chapter 5 to the multi-objective case.

Note that the fitness functions f and f' compute the same objective vector if s describes a spanning tree. This implies that the Pareto fronts for a given connected graph G contain the same objective vectors. As both fitness functions take only the weight vectors on the edges into account if s is a spanning tree, the Pareto sets of f and f' consist of all Pareto optimal spanning trees.

When considering a spanning tree T, we can create another spanning tree T' by integrating an edge $e \in E\backslash T$ into T and removing one edge of the created cycle $Cyc(T, e)$. Using such local changes we can transform a spanning tree T into another spanning tree S. The properties of such local changes have been examined in detail in Section 5.2. We will also make use of these results to obtain upper bounds on the runtime of simple MOEAs for the multi-objective minimum spanning tree problem.

10.2 Extremal Points of the Convex Hull

Let F be the Pareto front of a given instance. If we consider the bi-objective problem, the convex hull of F, denoted by $conv(F)$, is a piecewise linear

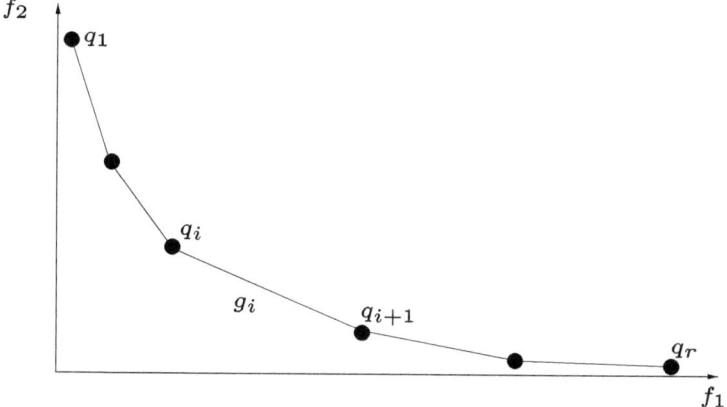

Fig. 10.1. The convex hull of the Pareto front F

function (see Figure 10.1). Note that for each spanning tree T on the convex hull there is a $\lambda \in [0, 1]$ such that T is a minimum spanning tree with respect to the single weight function $\lambda w_1 + (1 - \lambda)w_2$ (Knowles and Corne, 2001). We will use this in Section 10.3 to transform an arbitrary spanning tree S into a desired Pareto optimal spanning tree T on $conv(F)$ using Theorem 5.1.

Let q_1 and q_r be the Pareto optimal objective vectors with minimal weight with respect to f_1 and f_2, respectively. We denote by g_i, $1 \leq i \leq r - 1$, the linear functions with gradients m_i describing $conv(F)$. Then $i < j$ holds for two linear functions g_i and g_j iff $m_i < m_j$. Hence, the linear functions are ordered with respect to their increasing gradients. Let $q_i = (x_i, y_i)$, $2 \leq i \leq r - 1$, be the intersecting point of g_{i-1} and g_i. Our aim is to analyze the expected time until a simple MOEA has produced a population that includes a spanning tree for each vector of $F' = \{q_1, q_2, \ldots, q_r\}$. We call the vectors of F' the extremal points of the Pareto front.

The general idea of evolutionary algorithms is to create good approximations of optimal solutions for a given task. In the case of multi-objective problems, the task is to approximate the Pareto front.

Definition 10.1. *A solution x is called a c-approximation of a solution x^* if $f_1(x) \leq c \cdot f_1(x^*)$ and $f_2(x) \leq c \cdot f_2(x^*)$ holds. We also call the vector $(f_1(x), f_2(x))$ a c-approximation of the vector $(f_1(x^*), f_2(x^*))$ in this case. A set P of solutions is called a c-approximation of the Pareto front F if there exists for each solution x^* of the Pareto set a solution x in P that is a c-approximation of x^*.*

In the case of the minimization of two arbitrary functions with positive function values, the extremal points of the Pareto front give a 2-approximation of the Pareto front. This is shown in the following theorem.

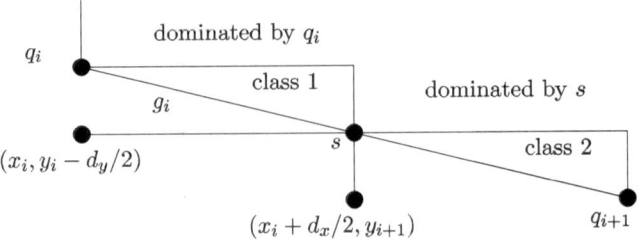

Fig. 10.2. Possible Pareto optimal vectors between two extremal points q_i and q_{i+1}

Theorem 10.2. *In the minimization of two objective functions with positive objective values, a set P containing for each extremal point a solution is a 2-approximation of the Pareto front.*

Proof. Consider the different possibilities for an arbitrary solution x^* of the Pareto set together with its Pareto optimal objective vector $q = (q_1, q_2)$, which is not an extremal point. Assume that $x_i < q_1 < x_{i+1}$ holds for some $i \in \{1, \ldots, r - 1\}$. This has to be the case because otherwise q would be dominated or the extremal points are not Pareto optimal. The situation for the two extremal points q_i and q_{i+1} is shown in Figure 10.2. If $q_2 \geq y_i$, q is not Pareto optimal as it is dominated by q_i. Let $d_x = x_{i+1} - x_i$ and $d_y = y_{i+1} - y_i$. Consider the vector $s = (s_1, s_2)$ with $s_1 = x_i + (d_x/2)$ and $s_2 = y_i - (d_y/2)$ in the objective space. As a 2-approximation of a vector q that dominates a vector q' is also a 2-approximation of q', it is not necessary to consider the vectors dominated by s. Note that s is a point on the linear function g_i which separates the possible vectors that have to be considered into two classes, Class 1 and Class 2. Class 1 includes all vectors q with $x_i < q_1 < x_i + d_x/2$ and $y_i - d_y/2 < q_2 < y_i$, and in addition the vector s. Class 2 includes all vectors q with $x_i + d_x/2 < q_1 < x_{i+1}$ and $y_{i+1} < q_2 < y_i - d_y/2$.

Clearly, $x_i \leq q_1$ and $y_{i+1} \leq q_2$. In the following, we show that if q belongs to Class 1, $y_i \leq 2 \cdot q_2$ holds and that if q belongs to Class 2, $x_{i+1} \leq 2 \cdot q_1$ holds. Hence, either q_i or q_{i+1} is a 2-approximation of q.

If q belongs to Class 1 then $q_1 \leq x_i + (d_x/2)$ holds. In this case, we get

$$q_2 \geq y_i + (q_1 - x_i) \cdot m_i$$
$$\geq y_i + (d_x/2) \cdot m_i$$
$$= y_i + (d_x/2)\frac{y_{i+1} - y_i}{x_{i+1} - x_i}$$
$$= y_i + (1/2)(y_{i+1} - y_i),$$

which implies that $2 \cdot q_2 \geq 2y_i + y_{i+1} - y_i \geq y_i$.

If q belongs to Class 2 then $q_2 < y_{i+1} - (d_y/2)$ holds. In this case, we get

$$q_1 \geq x_{i+1} + (q_2 - y_{i+1}) \cdot (1/m_i)$$
$$\geq x_{i+1} - (d_y/2) \cdot (1/m_i)$$
$$= x_{i+1} - (d_y/2)\frac{x_{i+1} - x_i}{y_{i+1} - y_i}$$
$$= x_{i+1} - (1/2)(x_{i+1} - x_i),$$

which implies that $2 \cdot q_1 \geq 2x_i + x_{i+1} - x_i \geq x_{i+1}$. $\qquad\square$

10.3 Analysis of GSEMO

In GSEMO, the first search point is chosen uniformly at random from the underlying search space. All results in this section hold for an arbitrary initial solution.

We start by analyzing GSEMO until it produces a population consisting of solutions which are connected graphs.

Lemma 10.3. *GSEMO working on the fitness function f or f' constructs a population consisting of connected graphs in expected time $O(m \log n)$.*

Proof. Due to the fitness functions, no steps increasing the number of connected components are accepted. The current population P consists at each time step of solutions having the same number of connected components, as otherwise the solution s with the smallest number of connected components would dominate a solution with a larger number of connected components in P. The decomposition of the objective space due to the number of connected components is shown in Figure 10.3. If P consists of search points with ℓ, $\ell \geq 2$, components, there are for each search point in P at least $\ell - 1$ edges whose inclusion decreases the number of connected components. The probability of a step decreasing the number of connected components is therefore at least $\frac{1}{e} \cdot \frac{\ell-1}{m}$, and its expected waiting time is bounded by $O(m/(\ell - 1))$. After we have decreased the number of connected components for one solution, all solutions with more connected components are deleted from the population. Hence, the expected time until the population consists only of solutions describing connected graphs is upper bounded by

$$em\left(1 + \ldots + \frac{1}{n-1}\right) = O(m \log n). \quad\square$$

Now we bound the expected time until P includes corresponding solutions for the objective vectors q_1 and q_r. Later, these solutions will serve as a basis for collecting solutions for the remaining extremal points.

Lemma 10.4. *GSEMO working on the fitness function f constructs a population that includes for each of the objective vectors q_1 and q_r a spanning tree in expected time $O(m^2 n w_{\min}(\log n + \log w_{\max}))$.*

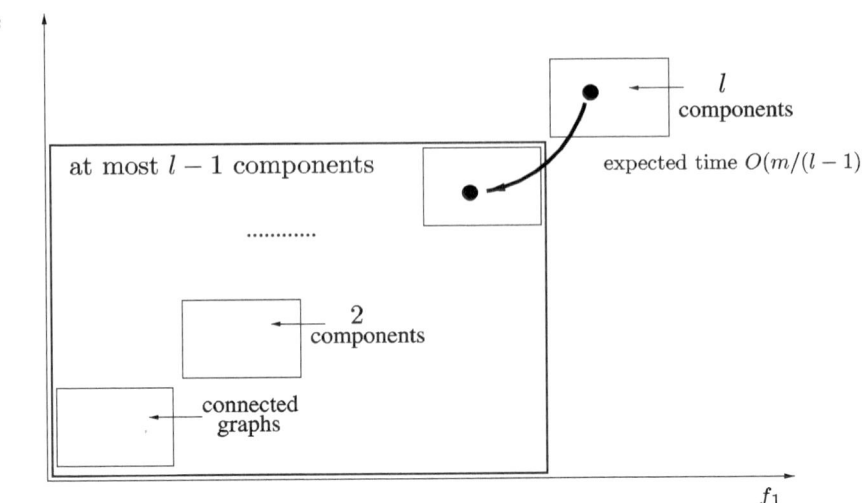

Fig. 10.3. Decomposition of the objective space due to the number of connected components

Proof. Using Lemma 10.3, we work under the assumption that P consists of individuals describing connected graphs. In this case, all individuals of P have the same number of edges. If there are N edges in each solution there are $N - (n-1)$ edges whose exclusion decreases the number of edges without increasing the number of connected components. Hence, the probability of decreasing the number of edges in the next step is at least $\frac{1}{e} \cdot \frac{N-(n-1)}{m}$, and we can bound the expected time to create a population consisting of spanning trees by

$$em\left(1 + \ldots + \frac{1}{m-(n-1)}\right) = O(m\log(m-n+1)) = O(m\log n).$$

If P consists of spanning trees, the population size is bounded from above by $(n-1)w_{\min}$ because there is only one spanning tree for each value of one single function in the population. We show an upper bound on the expected time to create a population including a spanning tree with vector q_1. The expected optimization time of $(1+1)$ EA$_b$ in the case of one cost function is bounded by $O(m^2(\log n + w_{\max}))$; see Theorem 5.7. We are working with a population of size $O(nw_{\min})$ and consider in each step the individual with the smallest weight with respect to the function w_1. In each step this individual is chosen with probability $\Omega\left(\frac{1}{nw_{\min}}\right)$. Following the ideas in the proof of Theorem 5.7, we can upper bound the expected time until P includes a spanning tree having minimal weight with respect to w_1 by $O(m^2nw_{\min}(\log n + \log w_1^{\max}))$.

It remains to bound the expected time to create, from a population with a minimal spanning tree S with respect to w_1, a population with a spanning tree T which is minimal with respect to w_1 and also Pareto optimal. If $|S \setminus T| = k$

holds, we can consider pairs of edges s_i, t_i with $s_i \in S \setminus T$ and $t_i \in T \setminus S$ due to the bijection given by Theorem 5.1. As S and T are both minimum spanning trees with respect to w_1, $w_1(s_i) = w_1(t_i)$ holds for $i = 1, \ldots k$ because otherwise we are able to improve T or S with respect to w_1. This contradicts the assumption that S and T are both minimum spanning trees with respect to w_1. $w_2(t_i) \leq w_2(s_i)$ holds for $i = 1, \ldots, k$ because otherwise we are able to improve T with respect to w_2 without changing the value of w_1 – a contradiction to the assumption that T is Pareto optimal. Hence, there are k exchange operations which turn S into T and the expected time to create T from S is bounded by $O(m^2 n w_{\min}(\log n + \log w_2^{\max}))$ using the ideas in the proof of Theorem 5.7.

Altogether we obtain an upper bound of $O(m^2 n w_{\min}(\log n + \log w_1^{\max} + \log w_2^{\max})) = O(m^2 n w_{\min}(\log n + \log w_{\max}))$ to construct a spanning tree with vector q_1. After we have constructed a population including a spanning tree for q_1, we can upper bound the expected time to create a population including for each of the vectors q_1 and q_r a spanning tree by $O(m^2 n w_{\min}(\log n + \log w_{\max}))$ using the same arguments as before, and this proves the lemma.   □

We give a similar bound for the fitness function f'. The main difference is that we can only guarantee a population size bounded by $O(m w_{\min})$.

Lemma 10.5. *GSEMO working on the fitness function f' constructs a population that includes for each of the objective vectors q_1 and q_r a spanning tree in expected time $O(m^3 w_{\min}(\log n + \log w_{\max}))$.*

Proof. We consider the expected time to create a spanning tree with vector q_1. At each time step, the population size is bounded by $m w_{\min}$, because there is only one search point for each value of one single function in the population. We consider in each step the connected graph with the minimal weight with respect to w_1 in P. Using the ideas of Lemma 10.4, a connected subgraph with minimal costs with respect to w_1 is constructed in expected time $O(m^3 w_{\min}(\log n + \log w_1^{\max}))$. This is a spanning tree because otherwise the weight of w_1 can be decreased. After that, we consider the spanning tree with minimal weight with respect to w_1 in P. We are in a position to minimize the weight of this spanning tree with respect to w_2 and this can be done in expected time $O(m^3 w_{\min}(\log n + \log w_2^{\max}))$ using the ideas of Lemma 10.4. The expected time to create a spanning tree with vector q_r can be bounded in the same way.   □

In the following, we work under the assumption that $F = F'$ holds, which means that the Pareto front consists only of extremal points. In this case we call the Pareto front strongly convex. Let $d(T, T') = |T \setminus T'|$ denote the distance of two spanning trees T and T', which equals the minimal number of exchanges of two edges for constructing T' from T.

Lemma 10.6. *Assume that the Pareto front F is strongly convex. For each spanning tree T with $w(T) = q_i$, $1 \leq i \leq r - 1$, there is a spanning tree T' with $w(T') = q_{i+1}$ and $d(T, T') = 1$.*

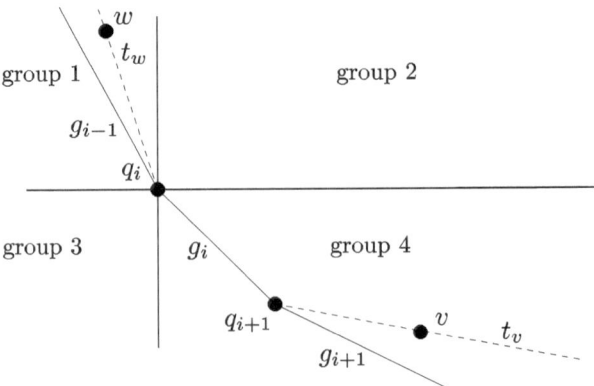

Fig. 10.4. The strongly convex Pareto front and the classification of exchange operations creating a spanning tree T' with vector q_{i+1} from a spanning tree T with vector q_i.

Proof. As T and T' are different, $d(T, T') > 0$ holds. We assume that T' is a spanning tree with vector q_{i+1}, which has minimal distance from T. Working under the assumption that $d(T, T') > 1$ holds for all spanning trees T' with vector q_{i+1}, we show a contradiction. We can apply Theorem 5.1 because for each spanning tree T' of the convex hull $conv(F)$ there is a $\lambda \in [0, 1]$ such that T' is a minimum spanning tree for the single weight function $\lambda w_1 + (1 - \lambda)w_2$. We partition the different exchange operations $exchange(e, e')$ inserting e and deleting e' due to Theorem 5.1 into four groups (see Figure 10.4). Let $d = exchange(e, e')$ and $w(d) = (w_1(e) - w_1(e'), w_2(e) - w_2(e'))$ be the vector describing the weight changes of this operation. d belongs to group 1 if $w_1(d) < 0$ and $w_2(d) > 0$, to group 2 if $w_1(d) \geq 0$ and $w_2(d) \geq 0$, to group 3 if $w_1(d) < 0$ and $w_2(d) < 0$, and to group 4 if $w_1(d) > 0$ and $w_2(d) < 0$.

There is no exchange operation d with $w(d) = (0, 0)$ because otherwise T' is not a spanning tree with vector q_{i+1} and minimal distance from T. All other operations belonging to group 2 are not possible because the remaining operations applied to T would construct a spanning tree dominating T' – a contradiction to the assumption that T' is Pareto optimal. Operations belonging to group 3 are not possible because they would construct a spanning tree dominating T. Let $q_i = (x_i, y_i)$, $1 \leq i \leq r$. There is no exchange operation belonging to group 4 which constructs a spanning tree T'' with value $x_i < w_1(T'') < x_{i+1}$ because q_{i+1} lexicographically follows q_i in the Pareto front. There is also no operation belonging to group 4 constructing a spanning tree with value $w_1(T'') \geq x_{i+1}$ and $w_2(T'') \geq y_{i+1}$ because otherwise the remaining operations applied to T construct a spanning tree which dominates T' – a contradiction to the assumption that T' is Pareto optimal.

Let M be the set of exchange operations constructing T' from T, $M_1 \subseteq M$ be the set of operations belonging to group 4, and $M_2 \subset M$ be the subset of operations belonging to group 1. Note that $M_1 \cup M_2 = M$ holds due to previous observations. We assume that M consists of more than one operation. As $x_{i+1} > x_i$ holds, M_1 is not empty. Let $v = (v_x, v_y)$ be the vector of the spanning tree constructed when all operations of M_1 are applied to T. $v_x > x_{i+1}$ and $v_y < y_{i+1}$ holds because otherwise we produce a spanning tree with vector q_{i+1} by one single operation (a contradiction to $d(T, T') > 1$), construct a spanning tree dominating T', or the remaining operations applied to T construct a spanning tree dominating T'. We consider the linear function t_v with the gradient m_v intersecting the points q_{i+1} and v. As F is strongly convex, $m_v \geq m_i > m_{i-1}$ holds. To construct a spanning tree with vector q_{i+1}, M_2 cannot be empty. Let $w = (w_x, w_y)$ be the vector of the spanning tree constructed when the operations of M_2 are applied to T and let t_w be the linear function with gradient m_w defined by q_i and w. As F is strongly convex, $m_w \leq m_{i-1}$ holds, which implies that $m_v > m_w$. Let $z = (z_x, z_y)$, $z_x < 0, z_y > 0$, be the vector such that $q_i + v + z = q_{i+1}$. As the operations of M applied to T construct T' with vector q_{i+1}, $w_x = z_x$ must hold. Taking the gradient m_w into account, we can compute the value of the second component as

$$v_y + m_w \cdot z_x > v_y + m_v \cdot z_x = y_{i+1}.$$

A contradiction to the assumption that the operations of M applied to T construct a spanning tree T' with vector q_{i+1}. Hence, T' has to be constructed from T by one single operation belonging to group 4.   □

Let $|F|$ be the number of Pareto optimal objective vectors. Note that $|F| \leq (n-1)w_{\min}$ holds. In the following, we show an upper bound on the expected time until the population P includes a corresponding spanning tree for each vector of a strongly convex Pareto front F.

Theorem 10.7. *The expected time until GSEMO working on the fitness function f or f' has constructed a population that includes a spanning tree for each vector of a strongly convex Pareto front F is bounded by $O(m^2 n w_{\min}(|F| + \log n + \log w_{\max}))$ or $O(m^3 w_{\min}(|F| + \log n + \log w_{\max}))$, respectively.*

Proof. We consider the fitness function f. Due to Lemma 10.4, the expected time to create a population including spanning trees for the Pareto optimal vectors q_1 and q_r is bounded by $O(m^2 n w_{\min}(\log n + \log w_{\max}))$. We assume that the population includes a spanning tree for each q_j, $1 \leq j \leq i$. For each spanning tree T with vector q_i, there exists a spanning tree T' with vector q_{i+1} and $d(T, T') = 1$. The probability of choosing the individual representing T in the next mutation step is at least $\frac{1}{(n-1)w_{\min}}$ because the population size is bounded by $(n-1)w_{\min}$. As $d(T, T') = 1$ holds for at least one spanning tree T' with vector q_{i+1}, the probability of constructing such a T', after having chosen the individual x describing T, is at least $\frac{1}{m^2}\left(1 - \frac{1}{m}\right)^{m-2} \geq \frac{1}{em^2}$. Hence, the expected time to create a spanning tree with vector q_{i+1} is bounded by $O(m^2 n w_{\min})$. As there are $|F|$ Pareto optimal vectors, the expected time until GSEMO constructs a spanning tree for

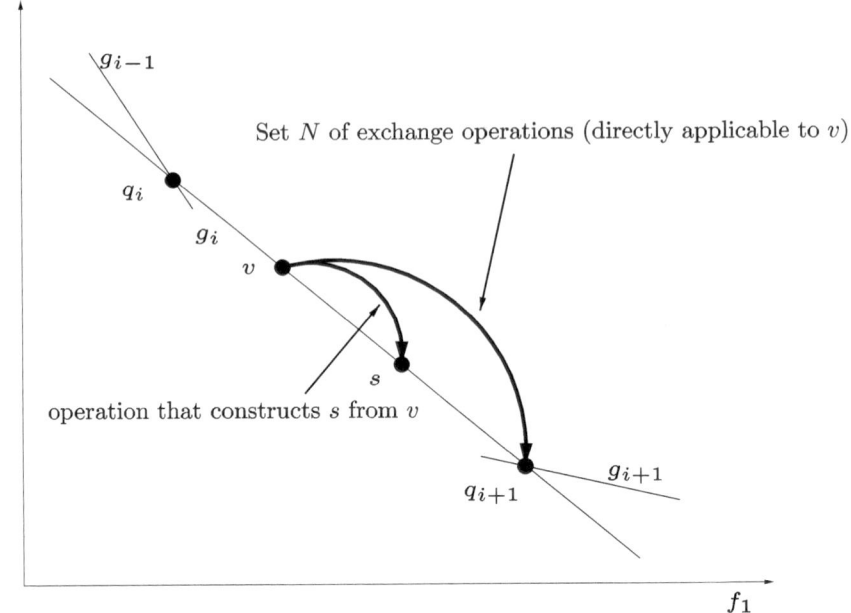

Fig. 10.5. Situation to compute the next extremal point

each Pareto optimal vector of a strongly convex Pareto front is bounded by $O(m^2 n w_{\min}(|F| + \log n + \log w_{\max}))$. The ideas can be easily adapted to f' using Lemma 10.5 and the upper bound $m w_{\min}$ on the population size. $\square$

We consider the general case now and give an upper bound on the expected time until GSEMO has constructed a population including a spanning tree for each extremal point $q \in F'$ of an arbitrary Pareto front F. Let $C = conv(F)$ be the set of objective vectors on the convex hull of F. Note that $|C| \leq (n-1) \cdot w_{\max}$ holds.

Theorem 10.8. *The expected time until GSEMO working on the fitness function f or f' has constructed a population that includes a spanning tree for each vector $q \in F'$ is bounded by $O(m^2 n w_{\min}(|C| + \log n + \log w_{\max}))$ or $O(m^3 w_{\min}(|C| + \log n + \log w_{\max}))$, respectively.*

Proof. Again, we consider the fitness function f and adapt the ideas to achieve the upper bound for f'. By Lemma 10.4, the population P includes spanning trees for the vectors q_1 and q_r after an expected number of $O(m^2 n w_{\min}(\log n + \log w_{\max}))$ steps. To transform a spanning tree of $conv(F)$ into another spanning tree of $conv(F)$, we use the set of exchange operations described by Theorem 5.1. Let T be a spanning tree with vector q_i, $1 \leq i \leq r-2$, and suppose that T' is a spanning tree with vector q_{i+1} and minimal distance to T. We denote by M the set of exchange operations classified as in the proof of Lemma 10.6 that construct T' from T. Using the arguments in the proof of Lemma 10.6, there are no exchanges belonging to group 2 or 3 in M. We show that each subset of M applied to T constructs a spanning tree on g_i. Suppose that a subset $M' \subseteq M$ of the operations constructs a spanning with a vector v not lying on g_i. This vector has to lie above g_i because otherwise it is outside of $conv(F)$. To construct a spanning tree with vector q_{i+1} on g_i, the operations of $M'' = M \setminus M'$ have to construct a spanning lying below g_i – a contradiction to the assumption that g_i is part of $conv(F)$.

We consider the spanning tree T'' with the lexicographic greatest vector $v = (v_x, v_y)$ on g_i in the population (see Figure 10.5). If $v \neq q_{i+1}$, T' can be constructed from T'' by a set N of exchanges of two edges, where each single exchange operation executed on T'' yields a spanning tree with a vector on g_i. As $v_x < x_{i+1}$ holds, there is at least one operation in this set N which constructs a spanning tree on g_i with vector $s = (s_x, s_y)$ where $v_x < s_x \leq x_{i+1}$ holds. Such a spanning tree is a spanning tree of $conv(F)$. Let C_i be the set of Pareto optimal vectors on g_i, $1 \leq i \leq r - 2$, excluding the lexicographic smallest vector and including the lexicographic greatest vector. The expected time to construct a population including spanning trees for the vectors of $\{q_1, \ldots q_i, q_{i+1}, q_r\}$ from a population P having spanning trees for the vectors of $\{q_1, \ldots, q_i, q_r\}$, $1 \leq i \leq r - 2$ is therefore upper bounded by $O(m^2 n w_{\min} |C_i|)$.

As $|C| = 1 + \sum_{i=1}^{r-2} |C_i|$ holds, the expected time, starting with a population including spanning trees for q_1 and q_r, to construct a population including a spanning tree for each vector of F' is bounded by $O(m^2 n w_{\min} |C|)$. Together with Lemma 10.4 we obtain the proposed bound.

To prove the upper bound for f', we use Lemma 10.5 and the upper bound of $m w_{\min}$ on the population size. Together with previous ideas we obtain an upper bound of $O(m^3 w_{\min} |C|)$ after constructing a population which includes spanning trees for q_1 and q_r and this proves the theorem. □

Conclusions

The multi-objective minimum spanning tree problem is one of the best-known multi-objective combinatorial optimization problems. We have analyzed evolutionary algorithms with respect to the expected time until they produce solutions of the Pareto front. In the case of a strongly convex Pareto front, we have achieved a pseudo-polynomial bound on the expected time until the population includes for each Pareto optimal objective vector a corresponding spanning tree. For an arbitrary Pareto front, we have considered the extremal points of the Pareto front. These points are of particular interest as they give a 2-approximation of the Pareto front. It has been shown that the population includes a solution for each extremal point after a pseudo-polynomial number of steps.

11

Minimum Spanning Trees Made Easier

In the previous chapter, we analyzed simple MOEAs on a given multi-objective optimization problem. In this chapter, we consider a multi-objective model of the minimum spanning tree problem. A single-objective model for the computation of minimum spanning trees has already been examined in Chapter 5. Our goal is to show that sometimes single-objective optimization problems can be solved much more easily by using a multi-objective model of the problem. This approach has opened a new research area in the field of multi-objective optimization and we want to discuss different results in the remainder of this book.

Sometimes, people try to turn multi-objective problems into single-objective ones, e.g., by optimizing a weighted sum of the fitness values of the single criteria. This may be useful in some applications but, in general, we do not obtain the information contained in the Pareto front and the corresponding search points. Many variants of evolutionary algorithms specialized for multi-objective optimization problems have been developed and applied, for a survey see the monographs of Deb (2001) and Coello Coello, Van Veldhuizen, and Lamont (2007). A conclusion from this discussion is that "multi-objective optimization is more (at least as) difficult than (as) single-objective optimization". This is true at least if the fitness values for the different criteria are "somehow independent". Without such an assumption, there is no reason to believe in the conclusion above.

The question arises about whether working in the more general framework of multi-objective optimization can lead to better understanding of a given problem or help us design more efficient algorithms for single-objective problems. Note that many single-objective problems have additional constraints that classify feasible and unfeasible solutions of the given search space. Such constraints can be relaxed such that additional objectives have to be optimized. Then the set of minimal elements contains the solution of the corresponding constrained single-objective problem. This has already been considered in the average case analysis of a well-known algorithm for the 0/1 knapsack problem. Beier and Vöcking (2004) have considered different input

F. Neumann, C. Witt, *Bioinspired Computation
in Combinatorial Optimization*, Natural Computing Series,
DOI 10.1007/978-3-642-16544-3_11, © Springer-Verlag Berlin Heidelberg 2010

distributions for this problem and shown that the number of minimal elements in the objective space is polynomially bounded. This implies that the well-known algorithm of Nemhauser and Ullman (1969) has an expected polynomial runtime for these distributions. A welcome by-product of a successful multi-objective approach is more information (a set of minimal elements instead of only one specific element) with even less computational effort.

We discuss the following scenario. The considered problem is a single-objective problem. It is possible to add some further criteria such that the Pareto front of the newly created multi-objective optimization problem is not too large and such that the solution of the multi-objective problem includes the solution of the single-objective problem. Solving the multi-objective problem instead of the single-objective problem implies computing the Pareto front instead of a single optimal value. Each considered search point contains more information than in the single-objective case since it contains also the fitness values for the additional criteria. At least in principle it is possible that this additional information improves the search behavior of evolutionary algorithms. This would imply that for solving difficult single-objective optimization problems one should also think about the possibility of modeling the problems as generalized multi-objective optimization problems.

The purpose of this chapter is to show that the considered scenario is not a fiction. We do not investigate artificial problems to support this claim but one of the combinatorial optimization problems contained in each textbook, namely the computation of minimum spanning trees. (Nobody should expect that evolutionary algorithms computing minimum spanning trees beat the well-known problem-specific algorithms.) In Chapter 5, we have already considered the runtime behavior of $\mathrm{RLS}_b^{1,2}$ and $(1+1)$ EA_b on this problem.

In Section 11.1, we introduce the two-objective variant of the minimum spanning tree problem which is the subject of our investigations and distinguish it from other multi-objective variants of the minimum spanning tree problem. In Section 11.2, we prove upper bounds on the expected optimization time of some evolutionary algorithms for multi-objective optimization applied to our problems. It turns out that they are asymptotically smaller than the lower bounds for the worst-case instances of simple evolutionary algorithms for the single-objective case. In order to investigate what happens for small problem dimensions and typical problem instances we present several experimental results in Section 11.3. We finish with some conclusions.

11.1 A Two-Objective Model

In Chapter 5, we have considered $\mathrm{RLS}_b^{1,2}$ and $(1+1)$ EA_b for the minimum spanning tree problem. We have penalized edge sets which do not describe connected graphs (and in one model additionally, edge sets containing cycles) and have shown the following results:

- The expected optimization time of $\mathrm{RLS}_b^{1,2}$ and $(1+1)$ EA_b is $O(m^2(\log n + \log w_{\max}))$ where $w_{\max}$ is the largest weight of the considered graph.
- There are graphs with n vertices, $m = \Theta(n^2)$ edges, and $w_{\max} = \Theta(n^2)$ such that the expected optimization time of $\mathrm{RLS}_b^{1,2}$ and $(1+1)$ EA_b equals $\Theta(m^2 \log n)$.

We discuss the reason for the expected optimization time of $\mathrm{RLS}_b^{1,2}$ and $(1+1)$ EA_b. If a search point describes a non-minimum spanning tree, 1-bit flips are not accepted. The new search point describes either an unconnected graph or a connected graph with a larger weight. We have to wait until a mutation step includes an edge and excludes a heavier one from the newly created cycle. The expected waiting time for a specified 2-bit flip equals $\Theta(m^2)$.

As already mentioned, the considered algorithms penalize the number of connected components. This motivates the following two-objective optimization model of the minimum spanning tree problem.

- The search space S equals $\{0,1\}^m$ for graphs on m edges and the search point s describes an edge set.
- The fitness function $f : S \to \mathbb{R}^2$ is defined by $f(s) = (c(s), w(s))$ where $c(s)$ is the number of connected components of the graph described by s and $w(s)$ is the total weight of all chosen edges.
- Both objectives have to be minimized.

We state some simple properties of this problem that are direct consequences of the presented model.

- The parameter $c(s)$ is an integer from $\{1, \ldots, n\}$.
- The first property implies that the populations of SEMO and GSEMO contain at most n search points and the Pareto front contains exactly n elements.
- The parameter $w(s)$ is an integer.

We have to be careful when discussing this model of the minimum spanning tree problem. In Chapter 10, we have discussed another type of multi-objective minimum spanning tree problem. Each edge has k different types of weights, i.e., $w(e) = (w_1(e), \ldots, w_k(e))$. Unconnected graphs are penalized, and the aim is to minimize $f(s)$ where s is not legal if s does not describe a connected graph, and $f(s)$ is the sum of all $w(e_i)$ where $s_i = 1$ otherwise. Similarly to other optimization problems, this multi-objective variant of a polynomially solvable problem is NP-hard.

11.2 Analysis of the Expected Optimization Time

The essential steps are 1-bit flips. In SEMO and GSEMO the initial search point is chosen uniformly at random from $\{0,1\}^m$. We discuss another possibility.

- The first search point is $s = 0^m$ describing the empty edge set. This is quite typical, e.g., for simulated annealing. We call the variants of SEMO and GSEMO using this initialization SEMO$_z$ and GSEMO$_z$, respectively.

Our analysis is simplified by knowing that P contains 0^m. Note that $f(0^m) = (n, 0)$ belongs to the Pareto front and 0^m is the only search point s with $c(s) = n$. First, we investigate the expected time until the population contains the empty edge set when starting with an arbitrary initial solution of the considered search space.

One might expect that we only have to wait until all edges of the initial search point s have been excluded. This is not true. It is possible that we accept the inclusion of edges since this decreases the number of connected components (although it increases the total weight). Later, we may exclude edges of the new search point s' without increasing the number of connected components. It is possible to construct a search point s'' which dominates s. Then s is eliminated and all search points in the population (perhaps only one) have more edges than s.

Hence, the situation is more complicated. Instead of the minimal number of edges of all search points in P, we analyze the minimal weight of all search points in P. One search point s^* with minimal weight has the largest number of connected components (otherwise, the search point s^{**} with $c(s^{**}) > c(s^*)$ is dominated by s^* and will be excluded from P). We analyze $w(s^*)$ and apply the method of the expected multiplicative distance decrease (see Section 4.2.3) and measure distance by $w(s^*)$. We have reached the aim of our investigations if $w(s^*) = 0$ since this implies $s^* = 0^m$. After initialization, $w(s^*) \leq W :=$ $w_1 + \cdots + w_m \leq m \cdot w_{max}$.

Lemma 11.1. *The expected time until the population of SEMO or GSEMO contains the empty edge set is $O(mn(\log n + \log w_{max}))$.*

Proof. We only investigate steps where the solution with minimal weight s^* is chosen for mutation. The probability of such a step is always at least $1/n$ since $|P| \leq n$. Hence, the expected time is only by a factor of at most n larger than the expected number of steps where s^* is chosen.

By renumbering, we may assume that s^* has chosen the first k edges. We investigate only steps flipping exactly one bit. This has probability 1 for SEMO and probability at least e^{-1} for GSEMO. These steps are accepted if they flip one of the first k edges. If the edge i is flipped, we obtain a search point whose weight is $w(s^*) - w_i$ and the minimal weight has been decreased by a factor of $1 - \frac{w_i}{w(s^*)}$. The average factor of the weight decrease equals

$$\frac{1}{m} \left(\sum_{1 \leq i \leq k} \left(1 - \frac{w_i}{w(s^*)}\right) + \sum_{k+1 \leq i \leq m} 1 \right) = 1 - \frac{1}{m}$$

if the choice of a non-existing edge is considered as a weight decrease by a factor of 1. The result $1 - \frac{1}{m}$ does not depend on the population. After

$M := \lceil (\ln 2) \cdot m \cdot (\log W + 1) \rceil$ steps choosing the current s^*, the expected weight of the new s^* is bounded above by $(1 - 1/m)^M \cdot W \leq \frac{1}{2}$. Applying Markov's inequality, the probability that $w(s^*) \geq 1$ is bounded above by $1/2$. Hence, $w(s^*) < 1$ holds with probability at least $1/2$. Since weights are integers, $w(s^*) < 1$ implies $w(s^*) = 0$. The expected number of phases of length M until $w(s^*) = 0$ is at most 2. Hence, altogether the expected waiting time for $s^* = 0^m$ is bounded above by $2 \cdot n \cdot M = O(mn(\log n + \log w_{max}))$ for SEMO. The corresponding value for GSEMO is larger at most by a factor of 3. $\square$

One may expect that the upper bound given in Theorem 11.1 is not exact for many graphs and starting points.

After having analyzed the expected time to produce a population that includes the empty edge set, we analyze to expected optimization under the condition that the empty edge set is included in the population.

Theorem 11.2. *The expected optimization time of SEMO$_z$ or GSEMO$_z$ is* $O(mn^2)$.

Proof. As long as the algorithm has not reached its goal, we consider the smallest i such that the population contains for each j, $i \leq j \leq n$, a Pareto optimal search point s_j with $f(s_j) = (j, w(s_j))$. This implies that the graph described by s_j consists of j connected components and has the minimal possible weight among all possible search points describing graphs with j connected components. After initialization, the population includes 0^m which has the smallest weight among all search points representing graphs with n connected components. Hence, i is well defined. The search point s_j is only excluded from the population if a search point s_j' with $f(s_j') = f(s_j)$ is included in the population. Hence, the crucial parameter i can only decrease and the search is successful if $i = 1$.

Finally, we investigate the probability of decreasing i. It is well known that a solution with $i - 1$ components and minimal weight can be constructed from a solution with i components and minimal weight by introducing the lightest edge that does not create a cycle. Therefore, it is sufficient to choose s_i for mutation (probability at least $1/n$) and to flip exactly one bit of a lightest edge connecting two components in the graph described by s_i (probability at least $1/m$ for SEMO$_z$ and at least $1/(em)$ for GSEMO$_z$). Hence, the expected waiting time to decrease the parameter i is bounded above by $O(nm)$. After at most $n - 1$ of such events the search is successful. $\square$

Corollary 11.3. *If the weights are bounded above by 2^n, SEMO and GSEMO find the Pareto front in the two-objective variant of the minimum spanning tree problem in an expected number of $O(mn^2)$ rounds.*

For dense graphs, this bound beats the bound $O(m^2 \log n)$ for the application of RLS$_b^{1,2}$ and $(1+1)$ EA$_b$ to the single-objective variant of the minimum spanning tree problem.

11.3 Experimental Results

The theoretical results are asymptotic ones. They reveal differences for worst-case instances and large m. We add experimental results that show what happens for typical instances and reasonable m. In order to compare randomized algorithms on perhaps randomly chosen instances, one may compare the average runtimes, but these values can be highly influenced by outliers. We have no hypothesis about the probability distribution describing the random runtime for constant input length. Hence, only parameter-free statistical tests can be applied. We apply the Mann-Whitney test (MWT) (Swinscow and Campbell, 2001) that ranks all observed runtimes. Small ranks correspond to small runtimes. If the average rank of the results of algorithm A_1 is smaller than that of A_2, MWT decides how likely it can be that such a difference or a larger one can occur under the assumption that A_1 is not more efficient than A_2. If the corresponding p-value is at most 0.05, we call the result significant, for 0.01 very significant, and for 0.001 highly significant. The statistical evaluation has been performed with the software SPSS. The tables contain the considered class of graphs, the average rank AR of different algorithms, and the p-value for the hypothesis that the algorithm with the smaller AR-value is likely to be faster.

The experiments consider the following graph classes.

- uniform$_n$: these are complete graphs with $m = \binom{n}{2}$ edges and the weights are chosen independently and uniformly at random from $\{1, \ldots, n\}$.
- uniformbd$_n$: each possible edge is chosen with probability $3/n$ leading to a small average degree of 3, unconnected graphs are rejected and the construction is repeated, and the weights of existing edges are chosen as for uniform$_n$.
- plane$_n$: the n vertices are placed randomly on the points of the two-dimensional grid $\{1, \ldots, n\} \times \{1, \ldots, n\}$, and the weight of an edge is the rounded Euclidean distance between the vertices.
- planebd$_n$: the n vertices are placed as for plane$_n$ but each edge is only considered with probability $3/n$ as for uniformbd$_n$.

These graph classes reflect different choices of weights (one non-metric and one metric) and the possibility of dense and sparse graphs. Our algorithms are RLS$_b^{1,2}$, (1+1) EA$_b$, SEMO, and GSEMO. The index z denotes the case where the initial search point is the empty edge set (or all-zero string). Without an index the initial search point is chosen uniformly at random. The runtimes of RLS$_b^{1,2}$ and (1+1) EA$_b$ denote the number of fitness evaluations until a minimum spanning tree is constructed. The runtimes of SEMO and GSEMO denote the number of rounds until, in one experiment, P contains a minimum spanning tree or until $f(P)$ equals the Pareto front. In each experiment, the compared algorithms are considered for 100 runs leading to an average rank of 100.5.

Table 11.1. Comparison of SEMO and GSEMO with different initial solutions until they have computed the Pareto front

Class	AR SEMO$_z$	AR SEMO	p-value	AR GSEMO$_z$	AR GSEMO	p-value
uniform$_{12}$	92.76	108.25	0.058	89.35	111.66	**0.006**
uniform$_{16}$	83.51	117.49	**< 0.001**	91.28	109.72	**0.024**
uniform$_{20}$	99.12	101.89	0.735	94.21	106.80	0.124
uniform$_{24}$	98.01	102.99	0.543	93.65	107.35	0.094
uniform$_{28}$	94.62	106.38	0.151	94.48	106.52	0.141
uniform$_{32}$	91.24	109.76	**0.024**	96.76	104.24	0.361
plane$_{12}$	81.61	119.39	**< 0.001**	88.14	112.86	**0.003**
plane$_{16}$	94.51	106.49	0.143	89.38	111.63	**0.007**
plane$_{20}$	97.17	103.83	0.416	95.15	105.85	0.191
plane$_{24}$	93.33	107.67	0.080	103.11	97.89	0.524
plane$_{28}$	90.58	110.43	**0.015**	93.09	107.91	0.070
plane$_{32}$	94.55	106.45	0.146	97.44	103.56	0.455

We analyze the influence of the initial search point. First, we consider the time until the Pareto front is computed. The results are shown in Table 11.1 and can be summarized as follows.

Result 1 *In 23 out of 24 experiments the variant starting with the empty edge set has the smaller AR-value. Only eight results are significant, among them five very significant and two of these highly significant.*

If we are only interested in the computation of a minimum spanning tree, we may expect that one sometimes computes a minimum spanning tree without computing the empty edge set. Indeed, the influence of the choice of the initial search point gets smaller. For the classes uniform$_n$, $n = 4i$ and $3 \leq i \leq 11$, there is no real difference between SEMO$_z$ and SEMO, while the AR-values of GSEMO are in eight of the nine experiments smaller than those for GSEMO$_z$. For the classes plane$_n$, $n = 4i$ and $3 \leq i \leq 11$, SEMO$_z$ beats SEMO (seven cases) and GSEMO$_z$ beats GSEMO (7 cases). We do not show the results in detail since they are not significant (with the exception of three out of 36 cases). The remaining experiments consider the more general case of an initial search point chosen uniformly at random.

We do not consider the worst-case instances for RLS$_b^{1,2}$ and (1+1) EA$_b$ presented in Chapter 3. This would be unfair to these algorithms. Nevertheless, the experiments of Briest, Brockhoff, Degener, Englert, Gunia, Heering, Jansen, Leifhelm, Plociennik, Roglin, Schweer, Sudholt, Tannenbaum, and Wegener (2004) indicate that, for n and m of reasonable size, dense random graphs are even harder than the asymptotic worst-case examples. This leads to the conjecture that SEMO beats RLS$_b^{1,2}$ and GSEMO beats its counterpart (1+1) EA$_b$. Here, the runtime measures the rounds until a minimum spanning tree is constructed. Table 11.2 shows that our conjecture holds for the

Table 11.2. Comparison of SEMO and GSEMO with their single-criteria counterparts on complete uniform and complete geometric instances

Class	AR $RLS_b^{1,2}$	AR SEMO	p-value	AR (1+1) EA_b	AR GSEMO	p-value
uniform$_{12}$	146.36	54.64	< **0.001**	147.79	53.32	< **0.001**
uniform$_{16}$	148.45	52.55	< **0.001**	149.28	51.72	< **0.001**
uniform$_{20}$	149.74	51.26	< **0.001**	149.40	51.60	< **0.001**
uniform$_{24}$	150.00	51.00	< **0.001**	150.29	50.71	< **0.001**
uniform$_{28}$	150.40	50.60	< **0.001**	150.23	50.77	< **0.001**
uniform$_{32}$	150.50	50.50	< **0.001**	150.50	50.50	< **0.001**
plane$_{12}$	141.43	59.58	< **0.001**	145.04	55.96	< **0.001**
plane$_{16}$	144.25	56.75	< **0.001**	148.28	52.72	< **0.001**
plane$_{20}$	149.47	51.53	< **0.001**	149.54	51.46	< **0.001**
plane$_{24}$	149.95	51.05	< **0.001**	149.89	51.11	< **0.001**
plane$_{28}$	150.40	50.60	< **0.001**	150.36	50.64	< **0.001**
plane$_{32}$	150.34	50.66	< **0.001**	150.28	50.72	< **0.001**

considered cases. Note that the average rank of 100 runs of one algorithm is at least 50.5. In several experiments, the AR-value of SEMO or GSEMO comes close to this value. For $n \geq 20$, all values are at most 51.6 and for small values of n the AR-values are smaller than 60. We can state the following result.

Result 2 *It is highly significant for all considered graph classes and graph sizes that SEMO outperforms $RLS_b^{1,2}$ and GSEMO outperforms (1+1) EA_b.*

The theoretical analysis of the algorithms gives values of $O(m^2 \log n)$ for $RLS_b^{1,2}$ and (1+1) EA_b and $O(mn^2)$ for SEMO and GSEMO (if the weights are reasonably bounded). For complete graphs, $m = \Theta(n^2)$ and we get values $n^4 \log n$ versus n^4. For sparse graphs, $m = \Theta(n)$ and we get values $n^2 \log n$ versus n^3. Although these are only upper bounds, one may expect different results for the sparse graphs from uniformbd$_n$ and planebd$_n$. Table 11.3 shows that this is indeed the case and we obtain the following result.

Result 3 *It is highly significant for* uniformbd$_n$ *and* $n \geq 24$ *and for* planebd$_n$ *and* $n \geq 16$ *(and the considered values of n) that* $RLS_b^{1,2}$ *outperforms SEMO. Similar results hold for (1+1) EA_b and GSEMO, but the results are highly significant only for large values of n, namely* $n \geq 32$, *for both graph classes.*

Note that the last group of experiments considers values of n up to 100.

Conclusions

It has been investigated whether the multi-objective variant of a single-variant optimization problem can lead to more efficient optimization processes. This is indeed the case for the well-known minimum spanning tree problem and

Table 11.3. Comparison of SEMO and GSEMO with their single-criteria counterparts on uniform and geometric instances with bounded average degree

Class	AR RLS$_b^{1,2}$	AR SEMO	p-value	AR (1+1) EA$_b$	AR GSEMO	p-value
uniformbd$_{12}$	91.91	109.09	**0.036**	101.44	99.57	0.819
uniformbd$_{16}$	90.62	110.39	**0.016**	103.54	97.46	0.458
uniformbd$_{20}$	89.79	111.22	**0.009**	98.98	102.02	0.710
uniformbd$_{24}$	73.19	127.82	**< 0.001**	91.53	109.47	**0.028**
uniformbd$_{28}$	78.01	122.99	**< 0.001**	93.03	107.98	0.068
uniformbd$_{32}$	77.92	123.08	**< 0.001**	80.85	120.15	**< 0.001**
uniformbd$_{40}$	73.02	127.98	**< 0.001**	84.37	116.63	**< 0.001**
uniformbd$_{60}$	65.40	135.60	**< 0.001**	71.22	129.78	**< 0.001**
uniformbd$_{80}$	56.70	144.30	**< 0.001**	58.72	142.28	**< 0.001**
uniformbd$_{100}$	54.99	146.01	**< 0.001**	58.47	142.53	**< 0.001**
planebd$_{12}$	97.56	103.45	0.472	105.24	95.77	0.247
planebd$_{16}$	81.88	119.13	**< 0.001**	96.79	104.22	0.364
planebd$_{20}$	81.06	119.95	**< 0.001**	101.70	99.30	0.769
planebd$_{24}$	84.45	116.55	**< 0.001**	86.52	114.48	**0.001**
planebd$_{28}$	81.94	119.06	**< 0.001**	88.45	112.55	**0.003**
planebd$_{32}$	71.53	129.47	**< 0.001**	80.86	120.14	**< 0.001**
planebd$_{40}$	67.18	133.82	**< 0.001**	74.57	126.44	**< 0.001**
planebd$_{60}$	56.59	144.41	**< 0.001**	60.69	140.31	**< 0.001**
planebd$_{80}$	52.98	148.02	**< 0.001**	59.60	141.40	**< 0.001**
planebd$_{100}$	52.21	148.79	**< 0.001**	52.30	148.70	**< 0.001**

randomly chosen dense graphs. For sparse connected graphs, it is better to use the single-objective variant of the problem. The results are obtained by a rigorous asymptotic analysis of the expected optimization time and by experiments on graphs of reasonable size.

We will see in the following chapters that a multi-objective approach for a single-objective optimization problem can also help with other problems. In particular, we examine *NP*-hard combinatorial optimization problems belonging to different areas of combinatorial optimization and show how a multi-objective approach can help us achieve better results than single-objective ones.

12

Covering Problems

In this chapter, we investigate the behavior of stochastic search algorithms on a class of covering problems. Such problems occur frequently in combinatorial optimization and it is therefore important to understand how stochastic search algorithms may deal with them. We will mainly consider the vertex cover problem, which is a well-known problem on graphs, but also extend our investigations to the much broader class of set covering problems. The goal is show the impact of different approaches that may be applied to covering problems.

In recent years, a number of publications regarding stochastic search algorithms for the vertex cover problem have appeared. First, some simple evolutionary algorithms for single-objective optimization have been investigated on this problem. It is shown in Friedrich, Hebbinghaus, Neumann, He, and Witt (2007) that a natural single-objective approach which minimizes the number of vertices and penalizes the number of uncovered edges has an exponential optimization even on simple bipartite graphs. Additional negative results regarding the single-objective search algorithms were presented by Oliveto, He, and Yao (2009) and Oliveto, He, and Yao (2008), who show that the use of populations in single-objective formulations does not necessarily allow for a significant increase in success probability. Based on these negative results, the combination of evolutionary algorithms and classical approximation algorithms was studied by Friedrich, He, Hebbinghaus, Neumann, and Witt (2009). The idea is to start with a solution produced by an approximation algorithm for the vertex cover problem and improve it over time by the stochastic search process of the evolutionary algorithm. The combination of evolutionary algorithms and different approximation algorithms is investigated and the benefits and limitations of this approach are pointed out. As a reaction to inherent worst-case assumptions, Witt (2009) studies the problem in random graphs and points out domains where a memetic local-search algorithm is efficient. We start this chapter with a presentation of the results by Friedrich, Hebbinghaus, Neumann, He, and Witt (2007). Later, extensions by

F. Neumann, C. Witt, *Bioinspired Computation*
in Combinatorial Optimization, Natural Computing Series,
DOI 10.1007/978-3-642-16544-3_12, © Springer-Verlag Berlin Heidelberg 2010

Oliveto et al. (2009) and an alternative view taken by Kratsch and Neumann (2009) are dealt with.

In Section 12.1, we introduce the problems investigated in this chapter. We start our analyses with investigations of (1+1) EA$_b$ and RLS in Section 12.2. After carrying out these investigations, we consider multi-objective models and prove that they can significantly help us come up with better stochastic search algorithms in Section 12.3. In this section we also present a characterization of the multi-objective models as so-called fixed-parameter algorithms.

12.1 Problem Formulation and Representation

We first investigate the *vertex cover* problem mentioned in Section 2.1 as a classical *NP*-hard problem. Recall that the input is given by an undirected graph $G = (V, E)$ and the task is to compute a minimum set of vertices $V' \subseteq V$ such that each for each edge e, $e \cap V' \neq \emptyset$, i.e., each edge contains at least one of the vertices in V'.

As we are looking for optimal subsets, we are able to proceed as in the previous chapters and to encode solutions as bitstrings. We consider the search space $\{0, 1\}^n$ where each bit x_i of a search point x corresponds to a vertex v_i of the given graph G. The vertex v_i is chosen in the solution x iff $x_i = 1$. The task is to find a solution x with a minimum number of vertices that covers all edges. This motivates us to introduce a fitness function which is based on the number of uncovered edges of x ($u(x)$) as well as the number of chosen vertices ($|x|_1$). Note that $u(x)$ may be used to direct the search process towards a feasible solution, i.e., a solution x for which $u(x) = 0$ holds. We consider the fitness function

$$f(x) = (u(x), |x|_1),$$

which takes into the account the number of uncovered edges and the number of chosen vertices. When considering (1+1) EA$_b$ our aim is to minimize f with respect to the lexicographic order, i.e., the main goal is to minimize the number of uncovered edges, which leads to solutions that are vertex covers. Afterwards, the number of chosen vertices is minimized.

Taking a multi-objective view on the problem, we do not give a preference to $u(x)$ or $|x|_1$. Instead, we treat these two objectives in the same way and optimize with respect to Pareto dominance. In the end, we pick the solution with $u(x) = 0$ and compare its number of chosen vertices to the number of vertices chosen in a minimum vertex cover.

The *set cover* problem generalizes vertex cover in the following way to weighted set systems. We are given a ground set $S = \{S_1, \ldots, S_m\}$ and a collection $C_1, \ldots, C_n$ of subsets of S with corresponding positive costs $c_1, \ldots, c_n$. We assume

$$\bigcup_{i=1}^{n} C_i = S,$$

i.e., S can be covered by the collection of subsets. Furthermore, we denote by $c_{\max} = \max_i c_i$ the maximum cost of a subset for a given instance. The goal is to find a minimum-cost selection $C_{i_1}, \ldots, C_{i_k}$, $1 \leq i_j \leq n$ and $1 \leq j \leq k$, of subsets such that all elements of S are covered. Note that the vertex cover problem for a given graph $G = (V, E)$ is a special set cover problem where $S = E$ and C_i denotes the set of edges incident to vertex v_i and $c_i = 1$ for $i \in \{1, \ldots, n\}$.

The set cover problem is, as a generalization of vertex cover, *NP*-hard. It cannot be approximated better than by a factor $\ln m$ unless certain assumptions from complexity theory do not hold (Feige, 1998; Raz and Safra, 1997). It is well known that Chvátal's simple greedy algorithm (Chvátal, 1979) achieves a worst-case approximation ratio of $O(\log m)$. In considering stochastic search algorithms for the search space $\{0, 1\}^n$, a search point $x \in \{0, 1\}^n$ encodes a selection of subsets. The function $p(x) = \sum_{i=1}^n c_i x_i$ measures the total cost of the selection and $u(x)$ denotes the number of elements of S that are uncovered. We use the same ideas as above and arrive at vector-valued fitness functions. Considering RLS_b^1 and $(1+1)$ EA_b for the set cover problem, the fitness of a search point x is given by the vector $f(x) = (u(x), p(x))$, which should be minimized with respect to the lexicographic order. In the multi-objective setting with the algorithms SEMO and GSEMO, we would like to minimize $u(x)$ and $p(x)$ at the same time and consider Pareto dominance.

12.2 Single-objective Optimization

12.2.1 The Vertex Cover problem

We start with the runtime behavior of $(1+1)$ EA_b and RLS_b^1 for the vertex cover problem. One result is that these algorithms are not able to achieve a good approximation even for bipartite graphs. Since the initial search point is drawn uniformly at random, it does not necessarily represent a valid vertex cover. In this case, the fitness function points the search towards such solutions.

Theorem 12.1. *The expected time until* RLS_b^1 *and (1+1)* EA_b *have produced a (not necessarily minimum) vertex cover is* $O(n \log n)$.

Proof. We prove the theorem for $(1+1)$ EA_b using the method of expected multiplicative distance decrease presented in Section 4.2.3. As the proof only works with 1-bit flips and all 1-bit flips are equally likely, the result also holds for RLS_b^1. Choosing all vertices is certainly a vertex cover and each vertex that has not been chosen before and is incident to an uncovered edge leads to an improvement with respect to the fitness function. Let k be the number of vertices that are incident to at least one uncovered edge. The number of uncovered edges is reduced from $u(x)$ to 0 by these k accepted 1-bit flips. As the prior aim is to minimize the number of uncovered edges, there are

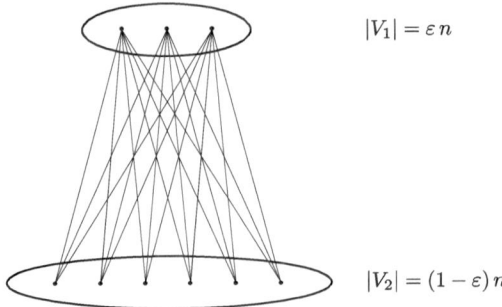

$|V_1| = \varepsilon n$

$|V_2| = (1-\varepsilon) n$

Fig. 12.1. The considered complete bipartite graph $B = (V, E)$ for $n = 9$ and $\varepsilon = \frac{1}{3}$

no accepted steps increasing the number of uncovered edges. Non-accepted 1-bit flips contribute a value of 0 to the reduction of the number of uncovered edges. We consider the expected decrease of $u(x)$ of an arbitrary 1-bit flip. Note that the probability of such steps is at least $(1/n)(1 - 1/n)^{n-1} \geq 1/e$. Choosing a 1-bit flip uniformly at random from among all 1-bit flips, the expected number of uncovered edges after this step is at most $(1-1/n) \cdot u(x)$, and after t steps this expected value is at most $(1 - 1/n)^t \cdot u(x)$. Choosing $t^* = cn \log n$, c an appropriate constant, this value is strictly less than $1/2$. As the number of uncovered edges is an integer, the probability of obtaining a vertex cover after t^* 1-bit flips is at least $1/2$ using Markov's inequality. This implies that the expected number of 1-bit flips to obtain a vertex cover is at most $2t^* = O(n \log n)$. The result follows as the probability of flipping a single bit in the next mutation step is at least $1/e$ and the expected waiting time for this event is therefore upper bounded by e. $\square$

Friedrich et al. (2007) prove that the previous upper bound is the best possible. They present the complete graph on n nodes as an instance where also a lower bound of $\Omega(n \log n)$ holds for the expected time until a minimum vertex cover is found. The proof goes back to the coupon collector's theorem; see Section 4.2.2. The interested reader is referred to the original work for more details.

In the following, we are dealing with the approximation quality of the single-objective search algorithms. The aim is to present an instance where RLS_b^1 and $(1+1)$ EA_b may get stuck with arbitrary bad approximations. This instance is a complete bipartite graph $B = (V, E)$, where $V = V_1 \cup V_2$ consists of two sets of non-equal size and the edge set $E = \{ \{v_i, v_j\} \mid v_i \in V_1 \wedge v_j \in V_2 \}$ consists of all edges that connect these two sets. W.l.o.g., we assume $|V_1| < |V_2|$. A minimum vertex cover is the set V_1 but both algorithms have a chance to determine the set V_2 as vertex cover. We consider the case $V_1 = \{v_1, v_2, \ldots, v_{\varepsilon n}\}$ and $V_2 = \{v_{\varepsilon n+1}, v_{\varepsilon n+2}, \ldots, v_n\}$, $0 < \varepsilon < 1/2$ and not necessarily constant.

The idea for the upcoming results is as follows. If RLS_b^1 has chosen all vertices of V_2 but some vertices of V_1 are missing, the algorithm cannot produce an approximation better than a factor of $(1 - \varepsilon)/\varepsilon$. On the graph B, the expected optimization time of RLS_b^1 is therefore infinite, as the next theorem shows.

Theorem 12.2. *With probability ε, RLS_b^1 cannot obtain an approximation better than a factor of $(1 - \varepsilon)/\varepsilon$ for B within a finite number of steps. In particular, the expected time to produce an approximation better than a factor of $(1 - \varepsilon)/\varepsilon$ on B is infinite.*

For the proof of Theorem 12.2, we will use the following lemma, which may be of independent interest.

Lemma 12.3. *A bin contains k red and ℓ blue balls. We take out the balls at random from the bin without replacement until there is either no red or no blue ball left. With probability $k/(\ell + k)$, there is no blue ball left, and with probability $\ell/(\ell + k)$, there is no red ball left.*

Proof. We take an alternative view on the model. Instead of taking out the balls until there is either no red or no blue ball left, we take out the balls at random from the bin without replacement until there is no ball left in the bin. The color of the last ball taken out of the bin clearly determines the ball color that has been removed at the time when there is only a single color left. Since every one of the $\binom{\ell+k}{k}$ orders of taking out all balls is equally likely and there are $\binom{\ell+k-1}{k}$ orders in which the last ball taken out is blue, the probability that the last ball is blue is

$$\binom{\ell + k - 1}{k} \Big/ \binom{\ell + k}{k} = \frac{(\ell + k - 1)! \, \ell! \, k!}{k! \, (\ell - 1)! \, (\ell + k)!} = \frac{\ell}{\ell + k}.$$

This proves the lemma. □

We are now able to prove Theorem 12.2.

Proof (Theorem 12.2). In the phase until the larger or the smaller vertex set is chosen completely by RLS_b^1, only steps that increase the number of vertices are accepted. This is because a reduction of the number of vertices in this phase reduces also the number of covered edges and thus the fitness value. Moreover, if the larger vertex set is the vertex set that is first determined completely by RLS_b^1, there is no chance for RLS_b^1 to determine the optimal solution, since only steps that reduce the number of vertices in the smaller vertex set are accepted. In this situation, the optimization time is infinite. Therefore, we have to prove that this happens with positive probability.

For this purpose, we apply Lemma 12.3. This is possible since the uniform choice of the initial search point is equivalent to the following procedure: We first choose a $k \in \{0, 1, \ldots, n\}$ following the binomial distribution $B(n, 1/2)$.

In other words, we choose k with probability $\binom{n}{k}(1/2)^n$. Afterwards, we choose successively k of the n vertices without repetition. It is easy to verify that the obtained search point is uniform on $\{0,1\}^n$. Lemma 12.3 is now applied starting with the empty subgraph and identifying the choice of a vertex in the initial search point with the event that the corresponding ball is taken out of the urn. Therefore, the probability is ε that the larger set of vertices is the first set that is completely chosen by RLS_b^1. This proves the theorem. $\square$

Theorem 12.2 shows that the approximability of RLS_b^1 for the vertex cover problem can be arbitrarily bad. Choosing, e.g., $\varepsilon = 1/n$, leads to a graph where V_1 consists of one single vertex. In this case, RLS_b^1 does not obtain an approximation better than a factor of $n-1$ with probability $1/n$. Note that an approximation of almost that quality can be obtained for an arbitrary graph by choosing all vertices of the given input.

Now we consider the behavior of $(1+1)$ EA_b on the graph B. After obtaining the vertex set V_2 and discarding the set V_1, $(1+1)$ EA_b cannot obtain a better approximation ratio than $(1-\varepsilon)/\varepsilon$ without flipping at least εn bits. If ε is not too small, $(1+1)$ EA_b can only leave this local optimum in the next mutation step with a probability that is exponentially small. Therefore, the expected optimization time under the condition that such a solution is produced before the optimal solution is exponential. The following theorem shows that this can lead to almost arbitrarily bad approximation ratios of roughly $n^{1-\delta}$, $\delta > 0$ a constant.

Theorem 12.4. *Let $\delta > 0$ be a constant and $n^{\delta-1} \le \varepsilon < 1/2$. The expected optimization time of $(1+1)$ EA_b on B (with $|V_1| = \varepsilon n$ and $|V_2| = (1-\varepsilon)\,n$) is exponential. Moreover, the expected time to produce an approximation better than a factor of $(1-\varepsilon)/\varepsilon$ is exponential.*

Proof. We investigate a run of two phases. In the first phase, we examine the probability that a vertex cover including all vertices of V_2 with at least one vertex missing in V_1 is constructed. In the second phase, we give a lower bound for the probability that a local optimum is obtained by removing all vertices of V_1. This local optimum can only be left by including all vertices of V_1 and removing at least εn vertices of V_2.

The first phase consists of $12en \ln n$ mutation steps. First we prove that $(1+1)$ EA_b obtains a vertex cover including all vertices of V_2 within this phase with probability at least $1/4$. We consider only the effect of steps that flip exactly one bit in V_2 and no other bit; these steps are called *simple V_2-steps* in the following. The probability of a simple V_2-step is $((1-\varepsilon)n)(1/n)(1-1/n)^{n-1} \ge (1-\varepsilon)/e$. Thus, the average waiting time for such a simple V_2-step is at most $e/(1-\varepsilon)$. We apply Markov's inequality on the waiting time for $k(1-\varepsilon)/(2e)$ of such steps. Hence, with probability at least $1/2$, there are in k steps of $(1+1)$ EA_b at least $k(1-\varepsilon)/(2e)$ simple V_2-steps. Using $1-\epsilon \ge 1/2$, this means that the considered phase of $12en \ln n$ mutation steps contains with probability at least $1/2$ at least $3n \ln n$ such simple V_2-steps. Considering this number, we apply the method of expected multiplicative distance decrease in a more precise way than in Theorem 12.1, where distance denotes the number of uncovered edges.

Let N be the current number of uncovered edges. All simple V_2-steps that add a vertex of V_2 are accepted and the total distance decrease of these steps is N since choosing all vertices from V_2 is clearly a valid vertex cover. Simple V_2-steps removing vertices of V_2 contribute a distance decrease of 0. There are, altogether, $(1-\varepsilon)n$ simple V_2-steps. Thus, a simple V_2-step decreases the number of uncovered edges by an expected factor of $1-1/((1-\varepsilon)n) \leq 1-1/n$. Since $N \leq n^2$, the expected number of uncovered edges after t simple V_2-steps is at most $(1-1/n)^t \cdot n^2$. Assuming $3n \ln n$ such steps, the expected number of uncovered edges after the phase is at most $n^2(1-1/n)^{3n \ln n} \leq 1/n$, which is strictly less than $1/2$. Hence, using Markov's inequality and the bound $1/2$ on the probability of having enough simple V_2-steps, a cover is produced by means of simple V_2-steps with probability at least $(1/2) \cdot (1/2) = 1/4$ in this phase.

Now we prove a lower bound on the probability that after $12en \ln n$ steps of $(1+1)$ EA$_b$, at least one vertex of V_1 has not been chosen. This is exactly the case if $(1+1)$ EA$_b$ completely discovers V_2 before completely discovering V_1. We base the analysis on the assumption that $12en \ln n$ steps lead to a vertex cover including all vertices from V_2 and note that this assumption does not decrease the probability of an unchosen vertex from V_1. By Chernoff bounds, there are with probability $1 - 2^{-\Omega(\varepsilon n)} = 1 - 2^{-\Omega(n^\delta)}$ at least $|V_1|/3 = \varepsilon n/3 \geq n^\delta/3$ unchosen vertices in V_1 in the initial solution. The probability that after $12en \ln n$ mutation steps of $(1+1)$ EA$_b$, a single vertex is chosen at least once is $1-(1-1/n)^{12en \ln n}$. Thus, the probability that at least one of the initially not chosen vertices of V_1 is not chosen after $12en \ln n$ mutation steps of $(1+1)$ EA$_b$ is

$$1 - \left(1 - \left(1 - \frac{1}{n}\right)^{12en \ln n}\right)^{\frac{n^\delta}{3}} \geq 1 - \left(1 - \frac{1}{n^{13e}}\right)^{\frac{n^\delta}{3}}$$

$$\geq 1 - e^{-\frac{n^{\delta-13e}}{3}} \geq 1 - \frac{1}{1 + \frac{n^{\delta-13e}}{3}} = \frac{\frac{n^{\delta-13e}}{3}}{1+\frac{n^{\delta-13e}}{3}} \geq \frac{n^{\delta-13e}}{6}.$$

For this estimation we used the fact $e^x \leq 1/(1-x)$ for $x < 1$. Altogether, the probability that $(1+1)$ EA$_b$ chooses all vertices of V_2 before choosing all vertices of V_1 is bounded from below by $(n^{\delta-13e}/6) \cdot (1/4) = n^{\delta-13e}/24$. Hence, the probability is at least bounded by an inverse polynomial from below.

We consider a second phase of $n^{3/2}$ mutation steps and show that all vertices of V_1 are removed with probability at least $1/15$. Let us assume that we start this phase with all vertices of V_2 and all but one vertex of V_1 in the current solution. This is the worst case for our analysis. In this phase (all vertices of V_2 and some vertices of V_1 chosen) the only mutation steps accepted by $(1+1)$ EA$_b$ are the following. Either all missing vertices of V_1 are chosen and at least as many vertices of V_2 are removed ("bad event"), or all vertices of V_2 are kept and the number of vertices in V_1 is decreased or stays the same by adding and removing some vertices ("good event"). The former mutation step has a probability of at most n^{-k}, where k denotes the current number of missing vertices in V_1. For the latter kind of mutation steps we restrict ourselves to 1-bit flips reducing the number of vertices in V_1. The probability of such a mutation step is at least $(\varepsilon n - k)/(en) \geq 1/(en)$. For our calculations we take only those two kinds of mutation steps into account, the "good event" with probability at least $(\varepsilon n - k)/(en)$ and the "bad event" with probability at most n^{-k}, since all other accepted mutation steps reduce or preserve the number of vertices in V_1. The probability that the "good event" occurs before the "bad event" is at least $\frac{1}{en}/(\frac{1}{en} + n^{-k}) = 1 - e/(n^{k-1} + e)$. Thus, the probability that the vertices of V_1 were all removed by (1+1) EA$_b$ before the "bad event" occurs is at least

$$\prod_{k=1}^{\varepsilon n-1} \left(1 - \frac{e}{n^{k-1}+e}\right) \geq \frac{1}{1+e}\left(1 - \frac{e}{n}\right)^{\frac{n-1}{2}} \geq \frac{e^{-e/2}}{1+e} > \frac{1}{15}.$$

The expected waiting time for removing all vertices of V_1 by (1+1) EA$_b$ is $O(n \log n)$ and therefore all vertices of V_1 are removed within $n^{3/2}$ steps with probability $1 - o(1)$ using Markov's inequality (always assuming that the "bad event" does not occur during this phase). Hence, the probability that (1+1) EA$_b$ determines the local minimum V_2 as vertex cover is at least $(1 - o(1)) \cdot n^{\delta - 13e}/360$. But if the current solution is V_2, every accepted mutation step has to add all the vertices of V_1 (and remove at least $|V_1|$ vertices of V_2). This occurs with probability at most $n^{-\varepsilon n} = n^{-\Omega(n^\delta)}$. Thus, the expected time until an approximation better than a factor of $(1 - \varepsilon)/\varepsilon$ is determined is at least

$$(1 - o(1)) \cdot \frac{n^{\delta - 13e}}{360} \cdot n^{\Omega(n^\delta)} = n^{\Omega(n^\delta)}.$$

This proves the theorem.   □

The preceding theorem proves that also the simple (1+1) EA$_b$ gets stuck on the bipartite graph class with at least constant probability. Multi-start variants of the algorithms can improve the success probability both for (1+1) EA and RLS$_b^1$ drastically (Oliveto et al., 2009) while the straightforward use of populations in single-objective formulations does not necessarily allow for a significant increase in success probability (Oliveto et al., 2008). We do not go into these results but present an instance which builds upon the bipartite graph class we have just considered. This instance is very difficult for the search algorithms in two respects. First, the probability of finding an optimum is so small that even multistart variants fail with overwhelming probability. Second, the approximation ratio where the search algorithms are stuck is $2 - o(1)$, i.e., roughly the approximation ratio guaranteed by the best problem-specific algorithms for the vertex cover problem (Karakostas, 2005).

Fig. 12.2. Graph $\tilde{B}$ consisting of independent copies of the bipartite instance from Figure 12.1

The following instance has, in essence, been proposed and investigated by Oliveto et al. (2009). We present a slightly simplified variant, which will be called $\tilde{B}$ in the following. This graph consists of $N := \sqrt{n}$ independent copies of the graph B on N vertices each and $\epsilon := N^{-2/3} = n^{-1/3}$. More precisely, the vertex set of $\tilde{B}$ equals the disjoint union of the sets $V_1^{(k)}$ and $V_2^{(k)}$ for all $k \in \{1, \ldots, N\}$, where $|V_1^{(k)}| = \epsilon N$ and $|V_2^{(k)}| = (1 - \epsilon)N$. For each k, there is a complete bipartite graph on the sets $V_1^{(k)}$ and $V_2^{(k)}$; this subgraph is denoted by B_k in the following. The edge set of $\tilde{B}$ equals the union of the edges of the B_k; hence there are no edges between different subgraphs. See Figure 12.2 for an illustration.

It will be shown that at least RLS_b^1 needs with overwhelming probability exponential time to obtain a better approximation ratio than $2 - o(1)$. The underlying idea is to consider the situation after the algorithm has reached a local optimum. We will prove that each subgraph has a probability of at most $1 - \epsilon$ of being optimized, i.e., that only the ϵN vertices in the corresponding V_1 set are chosen. Hence, this happens for an expected number of at most $(1 - \epsilon)N$ of the N subgraphs. Otherwise, $(1 - \epsilon)N$ vertices are chosen to cover the subgraph. In expectation, this happens for at least ϵN subgraphs. The total expected number of vertices in the cover is then at least

$$\epsilon N \cdot (1 - \epsilon)N + (1 - \epsilon)N \cdot \epsilon N = 2\epsilon n - 2\epsilon^2 n = (1 - o(1)) \cdot 2\epsilon n$$

using $\epsilon = o(1)$ while an optimal vertex cover consists of only ϵn vertices. This corresponds to the desired approximation ratio $2 - o(1)$. Since the subgraphs are independent, Chernoff bounds can be applied such that an approximation ratio $2 - o(1)$ holds also with overwhelming probability.

We make the ideas precise for RLS_b^1.

Theorem 12.5. *With probability $1 - 2^{-\Omega(n^{1/12})}$, RLS_b^1 needs on $\tilde{B}$ an infinite number of steps in order to obtain a solution with an approximation ratio better than $2 - o(1)$.*

Proof. For each subgraph B_k, $1 \leq k \leq N$, the ideas of Theorem 12.2 are applied. Steps that flip a vertex belonging to B_k are called *relevant* with respect to B_k. As long as there are uncovered edges with respect to B_k, the relevant steps add vertices from the corresponding V_1 and V_2 sets of the subgraph in the manner analyzed in Theorem 12.2. Hence, when all edges of B_k are covered for the first time, the set $V_2^{(k)}$ has been completely chosen with probability exactly ϵ. This happens independently for each subgraph. By Chernoff bounds, the probability that this happens for less than $(1 - n^{-1/24}) \cdot \epsilon N$ subgraphs is $2^{-\Omega(n^{-1/12}\epsilon N)} = 2^{-\Omega(n^{1/12})}$ according to Chernoff bounds. Otherwise, at least

$$((1 - n^{-1/24}) \cdot \epsilon N) \cdot (1 - \epsilon)N + ((1 - \epsilon)N) \cdot \epsilon N = (1 - o(1)) \cdot 2\epsilon n$$

vertices are in the cover, corresponding to an approximation ratio of $2 - o(1)$. Since RLS_b^1 flips only one bit per step, the corresponding search point will never be improved. □

It would be nice to have a result in the flavor of Theorem 12.5 also for (1+1) EA_b. However, this would mean that we would have to bound the probability by at least ϵ that a subgraph ends up with the larger vertex set (the V_2-set) chosen. The proof of Theorem 12.4 reveals only a significantly smaller constant for this probability. One reason for this is that the balls-and-bins game cannot be applied in the same manner as for RLS_b^1. In fact, steps flipping several bits might change the game in favor of the smaller vertex set. Consider a 2-bit flip removing a vertex from V_2 and adding at the same time a vertex from V_1. Since vertices in V_1 have a comparatively large degree of $|V_2|$ (as opposed to the degree $|V_1|$ for the vertices from V_2), the number of uncovered edges might be decreased considerably by the considered step. The opposite step, removing a vertex from V_1 in favor of V_2, would be rejected in this case. Hence, the probability of (1+1) EA_b ending up with the set V_2 completely chosen seems to be lower than that for RLS_b^1.

Due to these difficulties, Oliveto et al. (2009) consider a modified (1+1) EA_b that starts not from a random search point but from the all-ones string, i.e., the vertex cover choosing all the vertices. This algorithm behaves as follows on the $\tilde{B}$ instance: Considering an arbitrary subgraph B_k, the first step relevant for B_k is studied. If this step flips only one bit of B_k, the probability of removing a vertex from $V_1^{(k)}$ equals ϵ. This wrong decision leads with probability $1 - o(1)$ to the removal of more vertices of $V_1^{(k)}$ until only $V_2^{(k)}$ covers the edges of B_k. To support this, the authors exploit the fact that a subgraph contains only $N = \sqrt{n}$ vertices. Hence, given that a step is relevant for B_k, the probability of flipping at least $j \geq 2$ vertices of B_k is at most $(\sqrt{n}/n)^{j-1} = (1/\sqrt{n})^{j-1}$. This means that most steps relevant for a subgraph flip only one vertex in it. Subgraphs where too many j-bit flips, $j \geq 2$, are observed, can be taken out of the consideration without spoiling the approximation ratio $2 - o(1)$. The final result is as follows.

Theorem 12.6. *With a probability exponentially close to 1, (1+1) EA_b initialized with the all-ones string needs on $\tilde{B}$ an exponential number of steps in order to obtain a solution with an approximation ratio better than $2 - o(1)$.*

12.2.2 The Set Cover problem

Finally, we generalize the negative results obtained for the graph B to the set cover problem. The idea is to consider subsets C_i, $1 \leq i \leq n$, that correspond to the set of edges incident to the different vertices of B and assign large costs to subsets corresponding to vertices in V_2 and small costs corresponding to vertices in V_1. We make this precise and denote our class of instances by C^*. Let

$$S = \{\{v_1, v_{\varepsilon n+1}\}, \ldots, \{v_1, v_n\},$$
$$\{v_2, v_{\varepsilon n+1}\}, \ldots, \{v_2, v_n\},$$
$$\ldots$$
$$\{v_{\varepsilon n}, v_{\varepsilon n+1}\}, \ldots, \{v_{\varepsilon n}, v_n\}\}$$

be the ground set,

$$C_i = \{\{v_i, v_{\varepsilon n+1}\}, \ldots, \{v_i, v_n\}\}$$

with $c_i = 1$, $1 \leq i \leq \varepsilon n$, and

$$C_k = \{\{v_k, v_1\}, \ldots, \{v_k, v_{\varepsilon n}\}\}$$

with $c_k = c_{\max}$, $\varepsilon n + 1 \leq k \leq n$, be the subsets with associated costs, where $c_{\max} \geq 1$. We assume that $c_{\max}$ is a large value (e.g., $c_{\max} = 2^n$) to show that the approximation achievable by RLS_b^1 and $(1+1)$ EA_b in expected polynomial time may be arbitrarily bad.

In the proofs of the Theorems 12.2 and 12.4, we examine the probability that RLS_b^1 and $(1+1)$ EA_b obtain the larger partition of the bipartite graph before the smaller one. As long as a vertex cover has not been obtained, each mutation step decreasing the number of uncovered edges is accepted. We can translate the arguments given in the proofs to the set cover instance. Vertices in the graph B are mapped to sets of C^*. Again, each mutation step reducing the number of uncovered elements is accepted and the probability of choosing the sets C_i, $1 \leq i \leq \varepsilon n$, before the sets C_j, $\varepsilon n + 1 \leq j \leq n$, can be bounded in the same way as for the graph B. Therefore, we can generalize the Theorems 12.2 and 12.4 to C^* in the following way.

Theorem 12.7. *With probability ε, RLS_b^1 cannot obtain an approximation better than a factor of $((1 - \varepsilon)c_{\max})/\varepsilon$ for C^* within a finite number of steps. Moreover, the expected time to produce an approximation better than a factor of $((1 - \varepsilon)c_{\max})/\varepsilon$ on C^* is infinite.*

Theorem 12.8. *Let $\delta > 0$ be a constant and $n^{\delta-1} \leq \varepsilon < 1/2$. The expected optimization time of $(1+1)$ EA_b on C^* is exponential. In particular, the expected time to produce an approximation better than a factor of $((1 - \varepsilon)c_{\max})/\varepsilon$ is exponential.*

Theorems 12.7 and 12.8 show that the approximation quality achievable in expected polynomial time can be made arbitrarily bad as long as $c_{\max}$ grows.

12.3 Multi-objective Optimization

12.3.1 General Results

In this section, we turn our view to the multi-objective approaches SEMO and GSEMO and investigate these on the vertex cover and the more general set cover problems. We also revisit the instances discussed in the last section, namely the bipartite graph B and its generalization to a set cover instance C^*. In contrast to the single-objective formulations, the multi-objective ones have the ability to find optimal solutions for these instances efficiently. The main reason for this is that the multi-objective model makes the algorithm behave more like a greedy algorithm. With regard to the instance B, each vertex of V_1 is incident on $(1 - \varepsilon)n$ edges while each vertex of V_2 is incident on εn edges. A greedy algorithm that starts with the empty vertex set and adds in each step a vertex which covers a largest number of edges uncovered up to now ends up with V_1 and produces therefore an optimal solution. Examples of such greedy algorithms for the vertex cover problem are discussed in Papadimitriou and Steiglitz (1998). We will comment on the relationship between greedy algorithms and multi-objective optimization in greater detail when the set cover problem is studied.

We start with the announced positive result for the simple bipartite graph instance.

Theorem 12.9. *The expected optimization time of SEMO and GSEMO on the bipartite graph B is $O(n^2 \log n)$.*

Proof. We prove the theorem for GSEMO. All subsets of V_1 are Pareto optimal. The objective vector of a subset $V' \subseteq V_1$ with $|V'| = k$ is $(m - k(1 - \varepsilon)n, k)$. The Pareto front contains the $|V_1| + 1 = \varepsilon n + 1$ objective vectors $(m, 0)$, $(m - (1 - \varepsilon)n, 1)$, $(m - 2(1 - \varepsilon)n, 2)$, ..., $(0, \varepsilon n)$, where $m = \varepsilon(1 - \varepsilon)n^2$. The population size is bounded by $O(n)$ as a population can never contain two individuals with equal number of vertices.

First, we determine the time until the Pareto optimal search point 0^n with value $(m, 0)$ is found. Since it is the only one with $|x|_1 = 0$, it is never removed from the population again. One way for GSEMO to get "closer" to $(m, 0)$ is to select the individual with the smallest $|x|_1$-value from the current population and mutate it so that the $|x|_1$-value decreases. By the Coupon Collector's theorem (see Section 4.2.2), this shows that $(m, 0)$ is included in the population after $O(n^2 \log n)$ steps with high probability since the population size is bounded by $O(n)$.

We now bound the time to discover the whole Pareto set after $(m, 0)$ is found. Since the probability of flipping a single bit in one step is at least $1/e$, the probability to get from one Pareto optimal solution $(m - k(1 - \varepsilon)n, k)$ to the "next" Pareto optimal solution $(m - (k+1)(1 - \varepsilon)n, k+1)$ is $(\varepsilon n - k)/(e n)$. Using again the linear size of the population, the expected number of steps to gain the whole Pareto front is at most $\sum_{k=0}^{\varepsilon n - 1}(e n^2)/(\varepsilon n - k) = O(n^2 \log n)$, which completes the proof. As only 1-bit flips are used in the proof, the result also holds for SEMO. □

We now turn our view to the set cover problem. In the following, we want to support our claim that a multi-objective model might be superior to a corresponding single-objective approach as it has the ability to simulate a greedy approach using partial solutions.

We start by showing that the expected optimization time of SEMO and GSEMO on C^* is polynomial. The following properties hold for the multi-objective model of the set cover problem. The all-zeros string is Pareto optimal since it covers no elements at zero cost. Moreover, any population of the multi-objective algorithms, which is a set of mutually non-dominating search points, can have at most $m + 1$ elements.

Theorem 12.10. *The expected optimization time of SEMO and GSEMO on C^* is $O(mn \, (\log c_{max} + \log n))$.*

Proof. To prove the theorem, we generalize some ideas already used in the proof of Theorem 12.9. The Pareto front consists of the objective vectors $(m, 0), (m - (1-\varepsilon) \, n, 1), (m - 2(1-\varepsilon) \, n, 2), (0, \varepsilon n)$, and a solution corresponding to the objective vector $(m - i(1 - \varepsilon) \, n, i)$, $1 \leq i \leq \varepsilon n$, chooses exactly i subsets from the set $\{C_1, \ldots, C_{\varepsilon n}\}$ of subsets with costs 1. We first consider the time until the search point 0^n with Pareto optimal objective vector $(m, 0)$ is included in the population.

To estimate this time, we consider the expected multiplicative decrease of the minimum p-value for the current population. The probability of choosing an individual with minimum p-value from among all individuals in the population is $\Omega(1/m)$ as the population size is bounded above by $m + 1$. Since flipping a single bit decreases the p-value by an expected factor of $1 - 1/(en)$ or better, the expected time until the all-zeros string is reached is bounded above by $O(mn \, (\log c_{max} + \log n))$.

After obtaining a Pareto optimal solution x with objective vector $(m - k(1-\varepsilon) \, n, k), 0 \leq k < \varepsilon n$, there are $\varepsilon n - k$ subsets of cost 1 that can be chosen to obtain a Pareto optimal solution whose objective vector is $(m - (k+1)(1 - \varepsilon) \, n, k+1)$. Taking into account the upper bound on the population size as well as flipping one of the desired bits in x, the probability that such a step happens in the next iteration is at least $(\varepsilon n - k)/(enm)$. Hence, the expected time to obtain for the "next" Pareto optimal objective vector a corresponding solution is upper bounded by $O((mn)/(\varepsilon n - k))$. Summing up over the different values of k, a solution for each Pareto optimal objective vector is produced after an expected number of $O(mn \log n)$ steps under the condition that the search point 0^n has been obtained before, which completes the proof. □

Up to now, we have pointed out classes of problems where the multi-objective approach achieves better approximations than the single-objective one. We have also shown that the single-objective algorithms can only achieve an almost trivial approximation ratio within an expected polynomial number of steps. In contrast to this we point out in the following that the multi-objective model leads to good approximations within an expected polynomial number of steps. Here, we are in particular interested in the expected number of steps until a solution x with $u(x) = 0$ has been produced that is a good approximation of an optimal one.

We will show that SEMO and GSEMO are able to efficiently find approximate solutions to arbitrary instances of the NP-hard set cover problem. As mentioned in Section 12.1, the following approximation quality is, up to a constant factor, the best we can hope for in polynomial time for arbitrary instances, unless $P = NP$ (Raz and Safra, 1997).

Theorem 12.11. *For any instance of the set cover problem and any initial search point, SEMO and GSEMO find an $(\ln(m)+1)$-approximate solution in an expected number of $O\left(m^2 n + mn\left(\log n + \log c_{\max}\right)\right)$ steps.*

Proof. The proof idea is to show that SEMO is able to proceed along the lines of the greedy algorithm for set cover introduced by Chvátal (1979); see also Vazirani (2001) for a detailed presentation. Let $H_m := \sum_{i=1}^{m} 1/i$ be the mth harmonic number and $R_k := H_m - H_{m-k}$, $0 \leq k \leq m$, the sum of the last k terms of H_m. While the greedy algorithm is able to find H_m-approximate solutions, SEMO creates R_k-approximate solutions that cover k elements for increasing values of k, i.e., it arrives at partial solutions that are at least as good as in the greedy algorithm. This procedure can be viewed as a kind of greedy algorithm based on the archive of non-dominating solutions. The expected time until the all-zeros string 0^n is reached is bounded above by $O(mn\left(\log c_{\max} + \log n\right))$ using the same ideas as those in the proof of Theorem 12.10.

Let OPT be the cost of an optimal solution. Let $c(x) = m - u(x)$ be the number of elements of S covered in a solution x. The remainder of the proof studies the so-called potential of the current population, which is the largest k such that there is an individual x in the population where $c(x) = k$ and $p(x) \leq R_k \cdot \text{OPT}$. The potential is well defined since now the all-zeros string is always in the population.

It is easy to see that the potential cannot decrease. We examine the expected time until the potential increases at least by 1. To this end, we apply the analysis of the greedy algorithm by Chvátal (1979) and use the notion of cost-effectiveness of a set, defined as the cost of the set divided by the number of newly covered elements. If there are $n - k$ elements left to cover and we add the most cost-effective set to cover some of these, all the newly covered elements are covered at a relative cost of at most $\text{OPT}/(n - k)$. Hence, if the cost of the selection was bounded above by $R_k \cdot \text{OPT}$ before and $k' \geq k+1$ elements are covered after the step, the cost is at most $R_{k'} \cdot \text{OPT}$ afterwards. The probability of choosing an individual that determines the current potential is bounded below by $\Omega(1/m)$. The probability of adding a most cost-effective set is bounded below by $1/(en)$ as it suffices to flip a certain bit. Since the potential can increase at most m times, the expected time is $O(m^2n)$ until an R_m-optimal, i.e., H_m-optimal, individual covering all elements is created.   □

12.3.2 Specialized Bounds for Vertex Cover

This section is based on a recent study by Kratsch and Neumann (2009) and takes a closer look at the vertex cover problem using multi-objective algorithms. In particular, we replace the $O(\log n)$ approximation ratio delivered by Theorem 12.11 with a ratio that is bounded by OPT. This improves upon the previous bound if OPT is sufficiently small. Moreover, by introducing speed-up techniques similar to those in Section 5.3.3, we can obtain a bound on the expected optimization time that is polynomial if OPT is assumed to be fixed. Runtime bounds of this kind are considered in the area of so-called *fixed-parameter tractable (FPT)* algorithms (see Downey and Fellows, 1999, for an introduction to this area), which motivates Kratsch and Neumann (2009) to characterize a variant of GSEMO as an *evolutionary FPT algorithm*.

In the following, we again consider GSEMO in the multi-objective formulation of the vertex cover problem. We have already explained that its population size is bounded from above by $n + 1$ and that the all-zeros string 0^n is reached after an expected number of $O(n^2 \log n)$ steps (see the proof of Theorem 12.9 for more details). Afterwards, the idea is to wait for GSEMO to create a partial solution that "approximates" a minimum vertex cover in certain respects. More precisely, for an arbitrary search point x, we define the residual graph $R(x)$ as the induced subgraph on the vertices that are *not* chosen by x; for example, $R(0^n)$ is the input graph G itself. Our partial solutions correspond to residual graphs with maximum degree OPT.

We bound the time until GSEMO obtains certain partial solutions. Later it will be shown how these solutions can be extended to valid vertex covers with approximation ratio OPT.

Lemma 12.12. *The expected number of steps of GSEMO until the current population contains for the first time a search point x satisfying the following two properties is bounded by $O(\text{OPT} \cdot n^4)$:*

1. *The vertices chosen by x form a subset of a minimum vertex cover,*
2. *the residual graph $R(x)$ has maximum degree OPT.*

Proof. We start our considerations at the first point of time where the all-zeros string 0^n is contained in the current population of GSEMO. The expected time for this is $O(n^2 \log n)$ and is covered by the bound $O(\text{OPT} \cdot n^4)$. In the remaining proof, the expected time until finding a so-called good search point, i.e., a search point satisfying the two properties, is bounded by $O(\text{OPT} \cdot n^4)$.

We denote by $L \subseteq V$ the set of vertices in the input graph that have a larger degree than OPT. Every optimal vertex cover must include all the vertices in L since otherwise the more than OPT neighbors of a vertex in L

vertices would have to be chosen. We assume $L \neq \emptyset$ since otherwise 0^n would be a good search point.

The expected time until a good search point is found is bounded by means of a potential function taking on at most $O(|E| \cdot \text{OPT})$ different values. We will prove that if there is no good search point in the current population, the potential is decreased in the next step with probability $\Omega(1/n^2)$, which implies the bound $O(n^2 \cdot |E| \cdot \text{OPT}) = O(\text{OPT} \cdot n^4)$ and, therefore, the lemma.

With respect to a population P, the value s_i, $1 \leq i \leq \text{OPT}$, denotes the smallest number of uncovered edges from among all search points x in P satisfying $|x|_1 \leq i$, i.e., search points choosing at most i vertices. The current potential of P is defined as the sum of its s_i-values.

Let a population P including 0^n be given and denote by k the largest index such that P contains search points $x_0, \ldots, x_k$ with objective vectors $(0, s_0), \ldots, (k, s_k)$ and selecting only vertices from L. This condition is trivially fulfilled for the point 0^n; hence k is well defined. If $k = \text{OPT}$ then $|L| \geq \text{OPT}$, which means that x_k is optimal since L is a subset of a vertex cover of size OPT. Hence, we assume $k < \text{OPT}$ since otherwise there is nothing to show. For the same reason, we assume that x_k does not constitute a good search point. Now the idea is to show that x_k can be mutated into a search point that is better in at least one s_i-value and, therefore, has a lower potential. The probability of the corresponding mutation will be bounded by $\Omega(1/n^2)$.

The definition of k ensures that x_k chooses only vertices from L, which, as mentioned above, are a subset of a minimum vertex cover. This is the first property of a good search point. Hence, since x_k is assumed not to be good, the second property must be violated, which means that the residual graph $R(x_k)$ has a vertex of degree at least $\text{OPT} + 1$. Let v be such a vertex. We distinguish between two cases:

Case 1: $s_k - s_{k+1} \leq \text{OPT}$. Then choosing x_k and flipping in v leads to a search point choosing $k+1$ vertices and leaving at most $s_k - (\text{OPT}+1) < s_{k+1}$ edges uncovered. This search point dominates the search point with objective vector $(k+1, s_{k+1})$, and, therefore, improves the s_{k+1}-value.

Case 2: $s_k - s_{k+1} > \text{OPT}$. Then P contains a search point x_{k+1} with objective vector $(k+1, s_{k+1})$ that, due to the definition of k, selects at least one vertex u outside L. Hence, u has degree at most OPT and flipping u out leads to a search point choosing k vertices and leaving at most $s_{k+1} + \text{OPT} < s_k$ edges uncovered. This search point dominates the search point with objective vector (k, s_k) and improves the s_k-value.

In both cases, we have identified a mutation improving the potential. Since this mutation changes only one bit and the population size is $O(n)$, its probability is $\Omega(1/n^2)$. This completes the proof. $\square$

The partial solutions studied in Lemma 12.12 are useful for approximating optimal vertex covers since the residual graphs of such solutions have a bounded maximum degree. More precisely, given such a partial solution x, there are at most $(\text{OPT} - |x|_1) \cdot \text{OPT}$ uncovered edges since $\text{OPT} - |x|_1$ vertices of degree at most OPT suffice to cover all of them. This will be exploited in the following theorem.

Theorem 12.13. *The expected number of steps until GSEMO finds an OPT-approximation to the vertex cover problem is bounded by* $O(\text{OPT} \cdot n^4)$.

Proof. We start our considerations at the first point of time where the current population contains a search point satisfying the conditions of Lemma 12.12. Let x be such a search point. Then $|x|_1 \leq \text{OPT}$ and the maximum degree of $R(x)$ is at most OPT. Hence, as mentioned above, the number of uncovered edges with respect to x satisfies $u(x) \leq (\text{OPT} - |x|_1) \cdot \text{OPT}$. This implies $|x|_1 + u(x) \leq \text{OPT}^2$. If x is dominated by any solution x' then $|x'|_1 + u(x') < |x|_1 + u(x) \leq \text{OPT}^2$. Hence, in all following steps, there is a search point y in the population such that $|y|_1 + u(y) \leq \text{OPT}^2$.

In the following, we again consider a potential function $u(P)$ for the current population P, defined as the minimum u-value among all search points x in P that satisfy $|x|_1 + u(x) \leq \text{OPT}^2$. We already know that the u-value and thus the potential are bounded from above by OPT^2 and note that the potential cannot increase in the run of GSEMO. Now let y be a search point that determines the current potential. If $u(y) = 0$ then $|y|_1 \leq \text{OPT}^2$, which means that y represents an OPT-approximation. Hence, it remains to estimate the number of steps until the potential drops to 0.

We consider still y, a search point determining the current potential. Assuming $u(y) > 0$, there is at least one vertex $v \in R(y)$ incident on an uncovered edge. Choosing y for mutation and flipping in v leads to a search point y' with $|y'|_1 + u(y') \leq |y|_1 + u(y) \leq \text{OPT}^2$ and $u(y') < u(y)$. By the choice of y, the new search point y' cannot be dominated by any search point in the current population; hence y' is accepted and leads to a population with decreased potential. Since the considered step has probability $\Omega(1/n^2)$ and the potential can take on at most OPT^2 values, the potential reaches 0 after at most $O(\text{OPT}^2 \cdot n^2) = O(\text{OPT} \cdot n^4)$ steps. $\square$

The previous theorem proves that GSEMO obtains an OPT-approximation in expected polynomial time, which is an important supplement to the $O(\log n)$ bound from Theorem 12.11 and yields a significant improvement if OPT is not too big. Although one cannot hope to obtain a general polynomial bound for the *NP*-hard vertex cover problem, it would also be nice to have a bound on the expected number of steps until an optimal vertex cover is produced. Trivially, this expected optimization time of GSEMO is bounded from above by $O(n^{\text{OPT}+1})$ since it is sufficient to wait for a step that flips OPT bits of the all-zeros string. This bound is polynomial only if OPT does not depend on n. However, it is well known that an optimal vertex cover can be found in time $2^{O(\text{OPT})}$ using the ideas of fixed-parameter algorithms (Downey and Fellows, 1999). Bounds of this kind cannot be proved for random search, the plain (1+1) EA, nor the original GSEMO. Nevertheless, Lemma 12.12

shows that GSEMO is "close" to such an optimization time bound. To make this clear, we introduce a different mutation operator that shares similarities with the speed-up techniques of Section 5.3.3. The new mutation operator is problem-specific and asymmetric since it favors flipping in vertices that are incident on uncovered edges.

Given a current search point $x \in \{0,1\}^n$ for the vertex cover problem, we now use the mutation operator described as Algorithm 23. GSEMO$_{\text{alt}}$ denotes GSEMO with this new mutation operator. It is easy to verify that also

---

Algorithm 23 Alternative mutation operator in GSEMO$_{\text{alt}}$

1. Let $U(x) \subseteq E$ denote the set of edges that are not covered by x and $S(x) \subseteq \{1, \ldots, n\}$ the vertices being incident on the edges in $U(x)$.
2. Choose $b \in \{0,1\}$ uniform at random.
3. If $b = 0$ or $S(x) = \emptyset$ flip each bit of x independently with probability $1/n$.
4. Otherwise flip each bit of $S(x)$ independently with probability $1/2$ and each other bit independently with probability $1/n$.

GSEMO$_{\text{alt}}$ reaches the all-zeros string in an expected number of $O(n^2 \log n)$ steps and that Lemma 12.12 also applies to GSEMO$_{\text{alt}}$. The reason is that GSEMO$_{\text{alt}}$ uses the standard mutation operator of GSEMO if $b = 0$, i.e., with probability at least $1/2$. Based on these prerequisites, we prove the following theorem.

Theorem 12.14. *The expected optimization time of GSEMO$_{\text{alt}}$ for the vertex cover problem is bounded by $O(\text{OPT} \cdot n^4 + n \cdot 2^{\text{OPT}+\text{OPT}^2})$.*

Proof. We start our considerations with a population containing a so-called good search point, i.e., one satisfying the two conditions of Lemma 12.12. The expected time until such a population is obtained appears as $O(\text{OPT} \cdot n^4)$ in the bound of the theorem.

Since a good search point is, by definition, a subset of a minimum vertex cover, we wait for a step that chooses a good search point and mutates the bits corresponding to the missing vertices from the minimum vertex cover. Let x be an arbitrary good search point. The residual graph $R(x)$ has maximum degree OPT and a vertex cover of size $\text{OPT} - |x|_1$. Each vertex in such a vertex cover can be adjacent to at most OPT non-isolated vertices, implying that $R(x)$ has at most $(\text{OPT} - |x|_1) + (\text{OPT} - |x|_1) \cdot \text{OPT} \le \text{OPT} + \text{OPT}^2$ non-isolated vertices. These are each independently flipped with probability $1/2$ if $b = 1$ holds in the alternative mutation operator. All $n' \le n - 1$ isolated vertices of $R(x)$, however, are not flipped with probability at least $(1 - 1/n)^{n-1} \ge 1/e$. Altogether, x is chosen and mutated into a minimum vertex cover with probability at least $\Omega((1/n) \cdot 2^{-\text{OPT}+\text{OPT}^2})$. Since this holds for any good search point, the expected time to obtain a minimum vertex cover from a good search point is at most $O(n \cdot 2^{\text{OPT}+\text{OPT}^2})$. $\square$

The previous theorem contains a runtime bound in the style of parameterized complexity and shows how stochastic search algorithms combined with a problem-specific component are able to follow the ideas of FPT algorithms. Kratsch and Neumann (2009) continue their research further in this direction by introducing a modified fitness function. Here the objective of minimizing the number of uncovered edges is replaced. Instead, the objective value of a linear program is minimized, where the linear program corresponds to a relaxation of an integer programming formulation of vertex cover. In this model, strong characterizations of fractional solutions to the problem can be exploited. Kratsch and Neumann (2009) show an improved approximation ratio of 2 for GSEMO and also an improved bound for GSEMO$_{\text{alt}}$ where the expected runtime is, in essence, dominated by the term $n \cdot 4^{\text{OPT}}$ instead of $n \cdot 2^{\text{OPT}+\text{OPT}^2}$. We refer the interested reader to the original work for further details.

Conclusions

In this section, we have considered the vertex cover and the more general set cover problems. Here the single-objective search algorithms $(1+1)$ EA$_b$ and RLS$_b$ are likely to get stuck at solutions whose approximation ratios are close to trivial. Multi-objective models represent a promising alternative. For the set cover problem, a simple GSEMO obtains the asymptotically best possible approximation ratio of $O(\log n)$. Moreover, using insights from the domain of parameterized complexity, an alternative approximation ratio bounded by OPT has been proved. Finally, by introducing an asymmetric mutation operator to GSEMO, bounds on the expected optimization time were obtained that match the requirements for fixed-parameter tractable algorithms. All analyses show that the archiving strategies of multi-objective algorithms enable the search to proceed in a structured way and also to make decisions that resemble thos of components of problem-specific greedy algorithms.

13

Cutting Problems

In this chapter, we consider another important class of problems belonging to the field of combinatorial optimization. We study cutting problems in a given weighted graph. The minimum s-t cut problem is one of the basic, classical problems in combinatorial optimization, operations research, and computer science (Cormen et al., 2001). Evolutionary algorithms have produced good results for various kinds of difficult cutting problems (Duarte, Sánchez, Fernández, and Cabido, 2005; Liang, Yao, Newton, and Hoffman, 2002; Puchinger, Raidl, and Koller, 2004).

The basic minimum s-t cut problem has the following formulation. We are given a connected directed graph $G = (V, E)$ on $n + 2$ vertices and m edges and a function $c : E \to \mathbb{N}_+$ that imposes positive integer costs on the edges. We denote by $c_{\max} = \max_{e \in E} c(e)$ the largest cost among all edges. Two nodes $s, t \in V$ are distinguished. We call s the source node and t the target node. An s-t cut $S \subseteq E$ is a set of edges such that there is no path from s to t when the edges of S are deleted from E. The cost of a subset of E is defined as the sum of the costs of its elements. In the minimum s-t cut problem, the goal is to find an s-t cut $S \subseteq E$ of minimum cost. The minimum s-t cut problem is highly related to the problem of computing a maximum flow from s to t. A flow in G is a vector in $\mathbb{R}^{|E|}$ (one component for each edge) such that:

1. $0 \le \mathit{flow}((u,v)) \le c((u,v))$ $\hspace{4cm}$ $\forall (u,v) \in E$
2. $\sum_{(u,v) \in E} \mathit{flow}((u,v)) = \sum_{(v,u) \in E} \mathit{flow}((v,u))$ $\hspace{1.5cm}$ $\forall v \in V \setminus \{s,t\}$

Here the function $c \colon E \to \mathbb{N}_+$ imposes capacity constraints for the flow that can be sent along each edge. The value of the flow from s to t in G (denoted by $|\mathit{flow}|$) is given by the value of the flow that leaves s, i.e.,

$$|\mathit{flow}| = \sum_{(s,u) \in E} \mathit{flow}((s,u)).$$

The maximum flow in a given directed graph is the maximal value of a flow that can be sent from s to t without violating the capacity constraints.

F. Neumann, C. Witt, *Bioinspired Computation*
in Combinatorial Optimization, Natural Computing Series,
DOI 10.1007/978-3-642-16544-3_13, © Springer-Verlag Berlin Heidelberg 2010

Due to the classical maximum flow minimum cut theorem (Papadimitriou and Steiglitz, 1998) the maximum flow from s to t in a given network equals the value of a minimum s-t cut.

The two mentioned basic problems can be solved in polynomial time. We consider different variants of evolutionary algorithms for this problem and present rigorous runtime results that are due to Neumann, Reichel, and Skutella (2008). Besides the classical minimum s-t cut problem, there are many other variants of cutting problems some of which are NP-hard. Examples are the maximum cut problem and the minimum multicut problem (see Korte and Vygen, 2005). The minimum multicut is a generalization of the minimum s-t cut problem. Instead of one source-sink pair, k source-sink pairs (s_i, t_i), $i = 1, \ldots, k$, are given and the goal is to find a set of edges of minimum cost that disconnects every sink t_i from its associated source s_i, $i = 1, \ldots, k$. We examine this problem as a generalization of the basic minimum s-t cut problem and present results based that have been obtained by Neumann and Reichel (2008).

We start by analyzing single-objective and multi-objective approaches for the minimum s-t cut problem. In Section 13.1, we investigate two simple single-objective approaches and present instances where they fail to achieve a minimum cut in polynomial time. In Section 13.2, we show that a multi-objective approach leads to a minimum s-t cut in expected polynomial time. Furthermore, our results show that this multi-objective approach computes in expected polynomial time to a factor k-approximation for the multicut problem consisting of k pairs of sink and terminal nodes.

13.1 Single-objective Approaches

We start by considering two single-objective models for the minimum s-t cut problem. The first one is node-based, the second is edge-based. In the node-based approach, we are searching for a partitioning of the vertices into two subsets, one containing s and the other containing t, such that the cost of the edges connecting the s to the t side of the cut is minimal. In the edge-based approach we search for a subset of edges of minimal cost such that the deletion of those edges disconnects t from s, i.e., the chosen edges constitute a cut.

13.1.1 Node-Based Approach

The minimum s-t cut problem consists of splitting the input graph into two components such that the cost of the edges crossing the partitions is minimal. Therefore, it seems natural to assign the vertices of $V \setminus \{s, t\}$ to either s or t such that the graph is split into two partitions S and T of vertices where $s \in S$ and $t \in T$ holds. Obviously, the edges leading from S to T constitute a cut and the goal is to minimize the cost of such a solution. The underlying search space is $\{0,1\}^n$, where $v_i \in S$ iff $x_i = 0$ and $v_i \in T$ iff $x_i = 1$, $1 \le i \le n$.

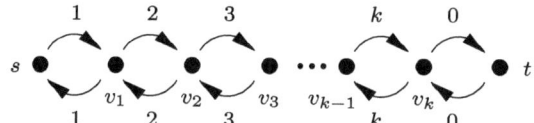

Fig. 13.1. Illustration of graph G_k

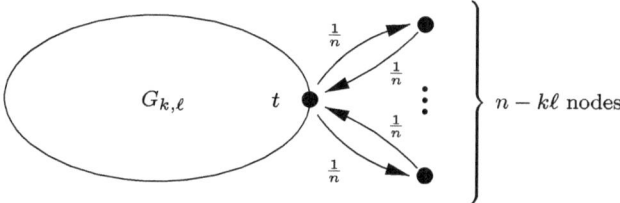

Fig. 13.2. Illustration of graph $G'_{k,\ell}$

The fitness of a search point x is given by

$$cost(x) = \sum_{e \in E \cap (S \times T)} c(e),$$

which computes the sum of the cost of all edges leading from S to T.

The node-based approach has a major drawback that has been pointed out in Neumann et al. (2008). The problem is that it imposes local optima with a large inferior neighborhood that make it hard to escape for stochastic search algorithms. We discuss such a class of instances in the following. To ease the presentation we use real-valued costs on the edges. However, an appropriate scaling can be used to come up with a corresponding class of instances that uses positive integer costs.

Consider the graph G_k (see Figure 13.1) given by a chain consisting of k interior vertices. Obviously, an optimal cut has cost 0 and assigns all vertices $V \setminus \{s, t\}$ to the source node s, i.e., the search point 0^k is the global optimum. However, the search of a single-objective algorithm such as RLS_b^1 and $(1+1)$ EA_b leads to a cut which assigns all vertices $V \setminus \{s, t\}$ to t, which constitutes a local optimum of cost 1, i.e., the search point 1^k is a local optimum. The probability of getting stuck in a local optimum becomes even higher if different copies of the graph G_k that share the vertices s and t are considered. Such a graph, called $G'_{k,\ell}$, is depicted in Figure 13.2 and contains a chain part consisting of ℓ copies of the graph G_k. In addition, it contains a star part consisting of $n - k\ell$ vertices that is directly connected t. Choosing $k = \Theta(n^{1/5})$ and $\ell = \Theta(n^{1/5})$, mutations affecting the chain part become unlikely in comparison to the clique part. In this way, the effect of mutation steps flipping more than one bit in the chain part can be controlled in a similar way to that in the proof of the lower bound of $(1+1)$ EA_b for the computation of a minimum spanning tree (see Theorem 5.9). Then it can be shown by investigating a typical run that RLS_b^1 and $(1+1)$ EA_b working on the fitness function *cost* compute with probability $1 - o(1)$ a local optimal solution which is not optimal with respect to at least one G_k in the chain part. We state the following theorem whose proof can be found in Neumann et al. (2008).

Theorem 13.1. *With probability $1 - o(1)$, the optimization time of RLS_b^1 and $(1+1)$ EA_b on $G'_{k,\ell}$ is $2^{\Omega(n^{1/10})}$.*

13.1.2 Edge-Based Approach

A solution to the minimum s-t cut problem is a set of edges. Therefore another natural approach is to work with a set of edges in stochastic search algorithms. The drawback of this approach is that not each edge set constitutes an s-t cut.

We now consider such an edge-based approach to obtain a minimum s-t cut. We work with bitstrings of length $m = |E|$ and consider RLS_b^1 and $(1+1)$ EA_b. Note, that $(1+1)$ EA_b flips each bit with probability $1/m$ in a mutation step.

As not every search point of the underlying search space represents a feasible solution, we have to penalize search points that do not represent an s-t cut. For a search point x, we do this by considering the value of a maximum flow that can be sent from s to t after taking out the chosen edges. Note that the flow value is 0 iff x represents a cut. Let $E(x) := \{e_i \in E \mid x_i = 1\}$ denote the subset of E corresponding to the 1s in a bitstring x. The fitness of a search point $x \in \{0, 1\}^m$ is given by

$$f(x) := cost(x) + \alpha \cdot flow(x)$$

for some $\alpha > 1$, where $cost(x) := \sum_{e \in E(x)} c(e)$ and $flow(x)$ denotes the maximum value of an s-t flow in the graph $G(x) := (V, E \setminus E(x))$. The capacity of an edge $e \in E$ equals its cost $c(e)$. The fitness function should be minimized.

Note that $flow(x)$ vanishes if and only if $E(x)$ contains an s-t cut of G. Hence, $flow(x)$ is a penalty term that penalizes bitstrings that do not correspond to a feasible solution. If $E(x)$ contains an s-t cut of G, the fitness function equals the value of the corresponding cut. A factor $\alpha \leq 1$ is unsuitable, since the empty set would have fitness smaller (or equal) than the global optimum.

It is well known that the value of a maximum flow in the graph is equal to the value of a minimum cut in the graph. However, considering just the value of a maximum flow, it is hard to gain structural information about the minimum cut. Therefore, it is interesting to examine whether stochastic search algorithms can take advantage of the value of a maximum flow in $G(x)$ for a given solution x.

In the following, we consider a class of graphs where simple stochastic search algorithms such RLS_b^1 and $(1+1)$ EA_b fail to obtain a minimum s-t cut when working with the introduced fitness function. Again, we show that

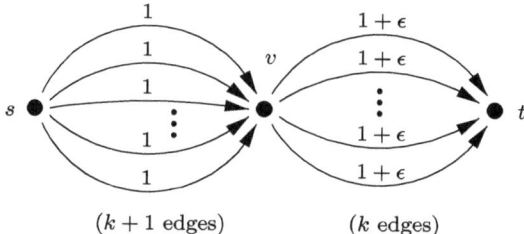

Fig. 13.3. Illustration of graph H_k

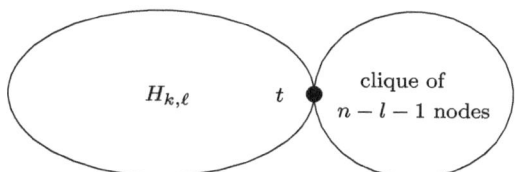

Fig. 13.4. Illustration of graph $H'_{k,\ell}$

the search space can contain local optima with a large inferior neighborhood which are produced by these algorithms before achieving an optimal solution. The instance is based on a graph H_k (see Figure 13.3), which consists of two vertices s, t and one interior vertex v. There are $k + 1$ edges of cost 1 leading form s to v and k edges of cost $1 + \epsilon$, where $1/k < \epsilon < 2/k$, leading from v to t. As $\epsilon > 1/k$ holds, all edges having weight 1 constitute a minimum s-t cut. The graph has the following property. If the number of chosen $(1 + \epsilon)$-edges is at some point of time larger than the number of chosen 1-edges by at least 2, then this property also holds for all later time steps if only 1-bit flips occur. This implies that if RLS_b^1 chooses an initial solution which satisfies the stated property, it will end up in the local optimum which consists of all $(1+\epsilon)$-edges.

The probability of getting stuck in a local optimum becomes even higher when different copies of the graph H_k that share the vertices s and t are considered. Such a graph, called $H'_{k,\ell}$, is depicted in Figure 13.4 and contains a bundle part consisting of ℓ copies of the graph H_k. In addition, it contains a clique part consisting of $n - \ell - 1$ vertices that is directly connected to t. All edges in the clique part have cost δ, where $0 < \delta \le (\alpha - 1)/n^2$, which implies that the influence of the edges in the clique part with respect to the fitness function is low. Choosing $\ell = \Theta(n^{1/10})$ and $k = \Theta(n^{4/10})$, bit flips in the bundle part become less likely than in the clique part. Again, a typical run investigating the effect of 1-bit flips can be considered in order to show that with probability $1 - o(1)$ a local optimum is reached which is not optimal with respect to at least one H_k. This leads to the following theorem, whose proof can be found in Neumann et al. (2008).

Theorem 13.2. *With probability* $1 - o(1)$*, the optimization time of* RLS_b^1 *and* $(1+1)$ EA_b *on* $H'_{k,\ell}$ *is* $2^{\Omega(n^{1/10})}$*.*

13.2 Multi-objective Model for the Multicut Problem

We have seen in the previous section that simple single-objective approaches may get stuck in local optima when dealing with the minimum s-t cut problem. This even holds if the fitness function takes into account the value of a maximum flow that can be sent from s to t by using the unchosen edges. We now consider a multi-objective approach for stochastic search algorithms to solve the minimum multi-cut which is a generalization of the minimum s-t cut problem. For the special case of the minimum s-t cut problem, we will show a polynomial upper bound on the expected time to compute a minimum cut and obtain approximation results for the general case.

The minimum multicut problem can be stated as follows. We are given a connected directed or undirected graph $G = (V, E)$ on n vertices and m edges and a cost function $c \colon E \to \mathbb{N}_+$ that imposes positive integer weights on the edges. Let $\{(s_1, t_1), \ldots, (s_k, t_k)\}$ be a set of k pairs with $s_i \neq t_i, 1 \leq i \leq k$. The source of commodity i is given by s_i, the target by t_i. We denote by $c_{\max} = \max_{e \in E} c(e)$ the largest cost among all edges.

A multicut $S \subseteq E$ is a set of edges such that there is no path from s_i to t_i in $(V, E \setminus S)$ for any commodity i. The cost of a subset of E is defined as the sum of the costs of its elements. The goal is to find a multicut $S \subseteq E$ of minimum cost.

To deal with the multicut problem we consider a generalization of the multi-objective approach to the multicut problem presented in Neumann and Reichel (2008). We want to consider a multi-objective model which takes into account the cost of a set of edges as well as the flow that can be sent through the network after the chosen edges have been deleted. Let F_i denote the value of a maximum s_i-t_i flow in G and define $F := \sum_{i=1}^k F_i$. We denote by F^* the sum of all flow values of a maximum multicommodity flow in G and by C^* the cost of a minimum multicut of G. Note that $F^* \leq C^* \leq C := m \cdot c_{\max}$. Furthermore, we have $F^* \leq F = \sum_{i=1}^k F_i \leq k \cdot F^* \leq k \cdot C^* \leq k \cdot C$.

We consider the fitness function $f \colon \{0, 1\}^m \to \mathbb{N}^2$, $f(x) = (cost(x), flow)$, where $cost(x) = \sum_{e \in E(x)} c(e)$, $flow(x) := \sum_{i=1}^k flow_i(x)$, and $flow_i(x)$ denotes the value of a maximum s_i-t_i flow in $G(x) := (V, E \setminus E(x))$.

Our goal is to show that the multi-objective model leads to an F/C^*-approximation for the multicut problem. Note that $F/C^* \leq k$; hence in the worst case we get a k-approximation. For the case $k = 1$, $F = C^*$ holds due to the maximum flow minimum cut theorem, which implies that a minimum s-t cut is obtained.

We denote by $L = \{x \in \{0, 1\}^m \mid cost(x) + flow(x) \leq F\}$ the set of search points whose objective vectors lie on or below the line given by the two objective values $(0, F)$ and $(F, 0)$. Figure 13.5 shows a graphical representation of the objective space for the general case.

The figure shows the case where the sequence F^*, C^*, F, $k \cdot F^*$, $k \cdot C^*$ is strictly increasing. Note that subsequent values may coincide and that C can be as small as C^*. Minimum multicuts x^* have objective vector $(C^*, 0)$;

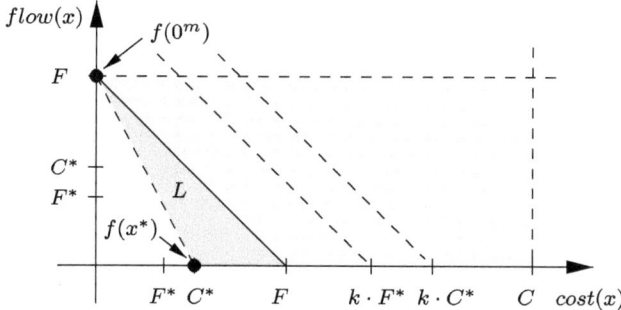

Fig. 13.5. Objective space of the fitness function $f(x) = (cost(x), flow(x))$ for the multicut problem

k-approximations lie on the the segment from $(C^*, 0)$ to $(\min\{k \cdot C^*, C\}, 0)$. The following lemma shows that the search points of L represent subsets of F/C^*-approximations of minimum multicuts.

Lemma 13.3. *Let $x \in L$. Then $E(x)$ is a subset of an F/C^*-approximation of a minimum multicut of G.*

Proof. Since $x \in L$ we have $cost(x) + flow(x) \leq F$. Let S denote a minimum multicut of $G(x)$. Then $E(x) \cup S$ is a multicut of G with $cost(E(x) \cup S) = cost(x) + cost(S)$. Since S is a minimum multicut of $G(x)$, its cost is not larger than the sum of the cost of the individual minimum s_i-t_i cuts, i.e., $cost(S) \leq flow(x)$. Hence, we have $cost(E(x) \cup S) \leq cost(x) + flow(x) \leq F \leq k \cdot F^* \leq k \cdot C^*$, which implies that $E(x) \cup S$ is an F/C^*-approximation of a minimum multicut of G. $\square$

Note that for $k = 1$ the set L is given by all search points x for which $cost(x) + flow(x) = F$ holds. This is an immediate consequence of the maximum flow minimum cut theorem. The preceding lemma implies the following condition for F/C^*-approximate solutions which will be essential for the analysis of the algorithms.

Corollary 13.4. *Let $x \in \{0, 1\}^m$ such that $flow(x) = 0$. Then $E(x)$ is an F/C^*-approximation of a minimum multicut of G if and only if $x \in L$.*

In the following, we examine how to obtain from a solution $x \in L$ with $flow(x) > 0$ another solution $x' \in L$ for which $flow(x') < flow(x)$. As a minimum multicut is a solution $z \in L$ for which $flow(z) - 0$; this is essential for our upper bound on the time to achieve an F/C^*-approximation of a minimum multicut. For $x \in \{0, 1\}^m$ and $e \in E$, let $x(e)$ be the value of bit corresponding to edge e in x. We define $x^{+e} \in \{0, 1\}^m$ by $x^{+e}(e) = 1$ and $x^{+e}(e') = x(e')$ for $e' \neq e$. We can bound $flow(x^{+e})$ in terms of $flow(x)$ as follows.

Lemma 13.5. *Let $x \in \{0,1\}^m$ and $e \in E$. Then $flow(x^{+e}) \geq flow(x) - kc(e)$.*

Proof. By the (single-commodity) maximum flow minimum cut theorem, we have

$$flow_i(x^{+e}) \geq flow_i(x) - c(e)$$

for each commodity i. Hence, we get

$$flow(x^{+e}) \geq \sum_{i=1}^{k}(flow_i(x) - c(e)) \geq flow(x) - kc(e). \quad \square$$

Now, we investigate how introducing an edge of a minimum s_i-t_i cut for some i in $G(x)$ changes the cost and flow value of a solution x.

Lemma 13.6. *Let $x \in \{0,1\}^m$ such that $flow_i(x) > 0$ for some commodity i. Let $e \in E \backslash E(x)$ be an edge of a minimum s_i-t_i cut of $G(x)$. Then $flow(x^{+e}) \leq flow(x) - c(e)$ and $cost(x^{+e}) + flow(x^{+e}) \leq cost(x) + flow(x)$.*

Proof. Since $flow_i(x) > 0$ the minimum s_i-t_i cut of $G(x)$ is not the empty set. Let $x \in E \setminus E(x)$ be an edge from such a minimum s_i-t_i cut. By the (single-commodity) maximum flow minimum cut theorem we have $flow_i(x^{+e}) = flow_i(x) - c(e)$. Furthermore, $flow_j(x^{+e}) \leq flow_j(x)$ holds for $j \neq i$. Summation over i yields the first claim.

Since $cost(x^{+e}) = cost(x) + c(e)$, the second claim follows directly from the first one. $\square$

Lemma 13.6 shows that we can make progress in L towards an F/C^*-approximation by choosing an edge $e \in G(x)$ that belongs to a minimum s_i-t_i cut in $G(x)$ for some i. The following corollary is an immediate consequence of the preceding lemma and the definition of L.

Corollary 13.7. *Let $x \in L$ a search point such that $flow(x) > 0$. Then there exists a 1-bit flip leading to a search point $x' \in L$ with $flow(x') < flow(x)$.*

After having examined some basic properties for the multi-objective model, we are now able to show runtime results for stochastic search algorithms First, we consider GSEMO and prove an upper bound on the expected time until this algorithm has achieved a F/C^*-approximation of the multicut problem.

Theorem 13.8. *The expected time until GSEMO working on the fitness function f constructs an F/C^*-approximation of a minimum multicut is $O(Fm(\log n + \log c_{\max}))$.*

Proof. The size of the population P is at most F as GSEMO keeps at each time step at most one solution per fixed flow value. First, we consider the time until $0^m \in L$ has been included into the population. Note that $cost(0^m) = 0$. Afterwards we study the time until $x \in L$ with $flow(x) = 0$ has been included.

Algorithm 24 DEMO (Diversity Evolutionary Multi-objective Optimizer)

1. Choose $x \in \{0,1\}^m$ uniformly at random.
2. Determine $f(x)$ and initialize $P := \{x\}$.
3. Repeat
   a) Choose $x \in P$ uniformly at random.
   b) Create an offspring y by flipping each bit of x independently with probability $1/m$.
   c) Let P unchanged if there is an $z \in P$ such that $b(z) \leq b(y)$ and $(b(z) \neq b(y)$ or $cost(z) + flow(z) < cost(y) + flow(y))$. Otherwise, exclude all z with $b(y) \leq b(z)$ and add y to P.

By Lemma 13.3, the edge set $E(x)$ is an F/C^*-approximation of a minimum multicut.

The expected time until GSEMO working on the fitness function f constructs 0^m is $O(Fm(\log n + \log c_{\max}))$. This can be proved using the technique of the expected multiplicative distance decrease where distance is measured with respect to $\min_{x \in P} cost(x)$.

Now we bound the time until a cut with the claimed approximation quality has been constructed. Once again we apply the method of the expected multiplicative distance decrease, now with respect to the flow value. Let x be the solution with the smallest flow value in $P \cap L$. Note that $\min_{x \in P \cap L} flow(x)$ does not increase during a run of GSEMO.

Consider a mutation step that selects x and performs an arbitrary 1-bit flip. Such a step is called a *good* step. The probability of a good step is lower bounded by $\Omega(1/F)$. By Lemma 13.3, $E(x)$ is a subset of an F/C^*-approximation of a minimum multicut, which can be obtained by including the remaining edges one by one. Therefore, a randomly chosen 1-bit flip decreases the minimum flow value in $P \cap L$ on average by a factor of at least $1 - 1/m$.

Using the method of the multiplicative distance decrease with respect to the flow value the expected time until $x' \in L$ with $flow(x') = 0$ has been discovered is $O(Fm(\log n + \log c_{\max}))$. $\quad\square$

We can state the following corollary for the minimum s-t cut problem.

Corollary 13.9. *If $k = 1$, the expected time until GSEMO working on the fitness function f constructs a minimum s-t cut is $O(Fm(\log n + \log c_{\max}))$.*

The upper bound on the expected optimization time of GSEMO is only polynomial if $c_{\max} = poly(n)$ holds. The reason for the pseudo-polynomial bound is that the population size can only be upper bounded by F. In fact, it is not too hard to come up with instances whose number of Pareto optimal objective vectors is exponential in the number of vertices. Due to this, the question arises about whether one has to keep for each Pareto optimal objective vector a corresponding individual in the population.

Dealing with a large Pareto front, multi-objective evolutionary algorithms usually do not keep for each nondominated objective vector one corresponding solution. Instead they work with a smaller population size and ensure diversity of the search points in the population with respect to the objective values.

One popular diversity mechanism that has been proposed in the literature is the so-called ϵ-dominance approach (see Laumanns, Thiele, Deb, and Zitzler, 2003). Here, objective vectors that are close to each other are grouped together and only one representative of such a group is kept.

We consider the DEMO algorithm (Diversity Evolutionary Multi-objective Optimizer) which differs from GSEMO by its partitioning of the objective space into boxes. A box includes objective vectors that are similar to each other and the algorithm keeps at most one individual per box in the population.

The objective space is partitioned into boxes by using the function

$$b\colon \{0,1\}^m \to \mathbb{N}^2$$

with

$$b_1(x) := \left\lfloor \frac{\log(1 + cost(x))}{\log(1 + \epsilon)} \right\rfloor$$

and

$$b_2(x) := \left\lfloor \frac{\log(1 + flow(x))}{\log(1 + \epsilon)} \right\rfloor,$$

where ϵ, $0 < \epsilon < 1$ is a parameter that determines the size of the boxes. This has the consequence that the population size of DEMO is upper bounded, as stated in the following lemma.

Lemma 13.10. *The population size $|P|$ of DEMO is upper bounded by*

$$B \le \frac{\log(1 + C)}{\log(1 + \epsilon)} = O(\epsilon^{-1} \log C) = O(\epsilon^{-1}(\log n + \log c_{\max})).$$

Proof. Since the $b_i(\cdot)$, $1 \le i \le 2$, value is a nonnegative integer and the population contains at most one search point per box, the population size is upper bounded by

$$\min\{\max_{x \in \{0,1\}^m} b_1(x), \max_{x \in \{0,1\}^m} b_2(x)\}.$$

Hence, we have $B \le \log(1 + C)/\log(1 + \epsilon) \le 2\log(1 + C)/\epsilon = O(\epsilon^{-1} \log C)$. $\qquad\square$

To obtain the upper bound on the runtime of DEMO, we first consider the time until the search point 0^m has been included in the population and analyze the time to achieve an F/C^*-approximation afterwards. DEMO does not keep all nondominated objective vectors found so far. The following lemma shows that for each search point $x \ne 0^m$ there is a 1-bit flip which produces from x a solution x' with a small b_1-value. Such operations will be essential for bounding the time until the solution 0^m has been included in the population.

Lemma 13.11. *Let $\epsilon \leq 1/m$ and $x \in \{0,1\}^m$ a search point such that $cost(x) > 0$. Then there exists a 1-bit flip leading to a search point $x' \in \{0,1\}^m$ with $b_1(x') < b_1(x)$.*

Proof. Consider all 1-bit flips that remove a single edge from $E(x)$. Among all resulting search points, consider a point x' that minimizes $y' := cost(x')$. Let $y := cost(x)$.

The repeated removal of edges in $E(x)$ yields the search point 0^m. Let $\ell := |E(x)| \leq m$. Since y' was minimal, $y' \leq (1 - \frac{1}{\ell})y$ holds. Since $\epsilon \leq \frac{1}{m} \leq \frac{1}{\ell}$ and $\ell \leq y$, we have

$$(1 + \epsilon)(1 + y') \leq 1 + \epsilon + (1 + \epsilon)\left(1 - \frac{1}{\ell}\right)y$$

$$\leq 1 + \frac{y}{\ell^2} + \left(1 + \frac{1}{\ell}\right)\left(1 - \frac{1}{\ell}\right)y = 1 + y.$$

This implies that

$$1 + \frac{\log(1 + y')}{\log(1 + \epsilon)} \leq \frac{\log(1 + y)}{\log(1 + \epsilon)}$$

and finally $b_1(x') < b_1(x)$. $\square$

Using Lemma 13.11, we now bound the expected time until DEMO has produced a population which includes the search point 0^m.

Lemma 13.12. *The expected time until DEMO working on the fitness function f includes the search point 0^m into the population is $O(m\epsilon^{-2}(\log^2 n + \log^2 c_{\max}))$.*

Proof. The archiving strategy of DEMO guarantees that whenever a non-empty box becomes empty, another search point whose box dominates the considered box is included into the population. Therefore, $\min_{x \in P} b_1(x)$ will never increase during the run of the algorithm.

Since the population size is bounded by B, the probability of picking a search point $x \in P$ with minimal b_1-value is $\Omega(1/B)$. By Lemma 13.11, there exists at least one 1-bit flip leading to a search point x' with $b_1(x') < b_1(x)$. The probability of generating such a search point x' is $\Omega(1/m)$. After at most B such steps, the b_1-value is zero, implying that we have found the search point 0^m. Hence, the expected time to include 0^m in the population is

$$O(B^2 m) = O(m\epsilon^{-2}\log^2 C) = O(m\epsilon^{-2}(\log^2 n + \log^2 c_{\max})).$$

This concludes the proof. $\square$

To come up with an upper bound for DEMO, it is necessary to examine how the algorithm progresses from a solution $x \in L$ to a solution of $x' \in L$ with $b_2(x') < b_2(x)$. The following lemma points out that this is possible by carrying out a special 1-bit flip.

Lemma 13.13. *Let $\epsilon \leq 1/m$ and $x \in L$ be a search point such that $flow(x) > 0$. Then there exists a 1-bit flip leading to a search point $x' \in L$ with $b_2(x') < b_2(x)$.*

Proof. By Lemma 13.7, there exists at least one 1-bit flip leading to a search point $x' \in L$ with $flow(x') < flow(x)$. Among all such search points, consider a point x' that minimizes $y' := flow(x')$. Let $y := flow(x)$.

The repeated application of Lemma 13.7 yields an F/C^*-approximation $E(x^*)$ of a minimum multicut of G. Let $\ell := |E(x^*)| - |E(x)| \leq m$. Since y' was minimal, $y' \leq (1 - \frac{1}{\ell})y$ holds. Since $\epsilon \leq \frac{1}{m} \leq \frac{1}{\ell}$ and $\ell \leq y$, we have $b_2(x') < b_2(x)$ by the same calculation as that in the proof of Lemma 13.11. □

Finally, we are able to prove the following theorem, which shows that the expected runtime of DEMO with an appropriate choice of ϵ is always polynomially bounded with respect to the given input.

Theorem 13.14. *Choosing $\epsilon \leq 1/m$, the expected time until DEMO working on the fitness function f constructs an F/C^*-approximation of a minimum multicut is $O(m\epsilon^{-2}(\log^2 n + \log^2 c_{\max}))$.*

Proof. Due to Lemma 13.12, the search point $0^m \in L$ has been included into the population after an expected number of $O(m\epsilon^{-2}(\log^2 n + \log^2 c_{\max}))$ steps. Hence, it is sufficient to consider the search process after having found a search point $x \in L$.

The archiving strategy of DEMO guarantees that whenever a non-empty box becomes empty, another search point whose box dominates the considered box is included into the population. Moreover, the tie-break rule ensures that a non-empty box with a search point $x \in P \cap L$ will never exchange that search point for a search point $x' \notin L$. Therefore, $\min_{x \in P \cap L} b_2(x)$ will never increase during the run of the algorithm.

Since the population size is bounded by B, the probability of picking a search point $x \in L$ with minimal b_2-value from among the search points in L is $\Omega(1/B)$. By Lemma 13.13, there exists at least one 1-bit flip leading to a search point $x' \in L$ with $b_2(x') < b_2(x)$. The probability of generating such a search point x' is $\Omega(1/m)$. After at most B such steps, the b_2-value is zero, implying that we have found a multicut. Since $x' \in L$, this multicut is an F/C^*-approximation of a minimum cut. Hence, the expected time to obtain an F/C^*-approximation of a minimum multicut is

$$O(B^2 m) = O(m\epsilon^{-2} \log^2 C) = O(m\epsilon^{-2}(\log^2 n + \log^2 c_{\max})).$$

This concludes the proof. □

For the minimum s-t cut problem, we get the following results as an immediate consequence of the previous theorem.

Corollary 13.15. *If $k = 1$ and $\epsilon \leq 1/m$, the expected time until DEMO working on the fitness function f constructs a minimum s-t cut is $O(m\epsilon^{-2}(\log^2 n + \log^2 c_{\max}))$.*

Conclusions

Finding minimum cuts in a given graph is one of the fundamental combinatorial optimization problems. We have examined how stochastic search algorithms can deal with this problem. Investigating two natural single-objective approaches, we have pointed out that they have to deal with local optima that have a large distance from the global one. This leads to exponential lower bounds on the runtime of RLS_b^1 and $(1+1)$ EA_b. Taking a multi-objective view on the problem, we have shown that GSEMO can solve the minimum s-t cut problem efficiently. To deal with a large number of trade-offs we have proposed using the ϵ-dominance approach leading to the algorithm DEMO. We have shown that this algorithm solves the minimum s-t cut problem as well as the generalized multicut problem in expected polynomial time.

The benefits and drawbacks for the use of the ϵ-dominance approach have been investigated in greater detail by Horoba and Neumann (2008). There are other diversity mechanisms that have been used in evolutionary multi-objective optimization, such as the density estimator or the use of the hypervolume indicator to direct the search. In Horoba and Neumann (2009), an approach using the density estimator is compared to the approach using ϵ-dominance as well as the approach of keeping all nondominated objective vectors. It is pointed out in which situations one mechanism is favored over the other. For the hypervolume indicator, the first runtime analysis was presented by Brockhoff, Friedrich, and Neumann (2008). We refer to the interested reader to the mentioned papers for further reading. For future research, it would be interesting to see how the other mentioned mechanisms can help us deal with combinatorial optimization problems that encounter an exponential number of trade-offs.

A

Appendix

We present some elementary mathematical material that is used throughout this book. Most of these basics in mathematics can be found in Feller (1968, 1971) and Motwani and Raghavan (1995).

A.1 Probability Distributions

Definition A.1 (Binomial distribution). *A random variable X follows the binomial distribution with parameters n and p if*

$$\mathrm{Prob}(X = k) \;=\; \binom{n}{k} \cdot p^k \cdot (1 - p)^{n-k}$$

for $k \in \{0, \ldots, n\}$. Its expectation is $E(X) = np$.

Illustratively, the random variable X counts the number of successes in n independent Bernoulli trials with probability p for a success.

Definition A.2 (Geometric distribution). *A random variable X follows the geometric distribution with parameter p if*

$$\mathrm{Prob}(X = k) \;=\; p^k \cdot (1 - p)$$

for $k \in \mathbb{N}_0$. Its expectation is $E(X) = 1/p$.

Illustratively, X counts the number of consecutive successes before the first failure in independent Bernoulli trials with success probability p.

Definition A.3 (Poisson distribution). *Let λ be a positive real number. A random variable X follows the Poisson distribution with parameter λ if*

$$\mathrm{Prob}(X = k) \;=\; \frac{\lambda^k e^{-\lambda}}{k!}$$

for $k \in \mathbb{N}_0$. Its expectation is $E(X) = \lambda$.

F. Neumann, C. Witt, *Bioinspired Computation*
in Combinatorial Optimization, Natural Computing Series,
DOI 10.1007/978-3-642-16544-3, © Springer-Verlag Berlin Heidelberg 2010

A.2 Deviation Inequalities

Proposition A.4 (Markov's inequality). *Let X be a random variable assuming only non-negative values. Then for all $t \in \mathbb{R}^+$,*

$$Prob(X \geq k \cdot E(X)) \leq 1/k.$$

Proposition A.5 (Chernoff bounds). *Let $X_1, X_2, \ldots, X_n$ be independent Poisson trials such that for $1 \leq i \leq n$ $Prob(X_i = 1) = p_i$, where $0 < p_i < 1$. Let $X = \sum_{i=1}^n X_i$, $\mu = E(X) = \sum_{i=1}^n p_i$. Then the following inequalities hold.*

$$Prob(X \geq (1+\delta)\mu) \leq \left(\frac{e^\delta}{(1+\delta)^{(1+\delta)}} \right)^\mu \qquad \delta > 0$$

$$Prob(X \geq (1+\delta)\mu) \leq e^{-\mu\delta^2/3} \qquad 0 < \delta \leq 1$$

$$Prob(X \leq (1-\delta)\mu) \leq e^{-\mu\delta^2/2} \qquad 0 < \delta \leq 1$$

Proposition A.6 (Chernoff-Hoeffding bound). *Let $X_1, \ldots, X_n$ be independent random variables such $a_i \leq X_i \leq b_i$ for $1 \leq i \leq n$. Denote $X = \sum_{i=1}^n X_i$. Then for any $\delta \geq 0$ the following inequalities hold.*

$$Prob(X \geq \mu + \delta) \leq e^{-2\delta^2 / \sum_{i=1}^n (b_i - a_i)^2}$$

$$Prob(X \leq \mu - \delta) \leq e^{-2\delta^2 / \sum_{i=1}^n (b_i - a_i)^2}$$

A.3 Other Useful Formulas

Proposition A.7 (Union bound). *For a finite or countably infinite sequence $A_1, A_2, A_3, \ldots,$ of events*

$$Prob\left(\cup_{i \geq 1} A_i\right) \leq \sum_{i \geq 1} Prob(A_i).$$

Proposition A.8 (Law of total probability). *For an event A and a partition of the sample space Ω into mutually disjoint events $B_1, \ldots, B_k$, i.e., $\cup_{i=1}^k B_i = \Omega$, it holds*

$$Prob(A) = \sum_{i=1}^k Prob(A \mid B_i) \cdot Prob(B_i),$$

where $\mathrm{Prob}(A \mid B_i)$ denotes the conditional probability of A, given B_i.

Proposition A.9 (Stirling's formula). *For any $n \in \mathbb{N}$*

$$\sqrt{2\pi n}\, n^n e^{-n} < n! < \sqrt{3\pi n}\, n^n e^{-n}$$

holds.

Proposition A.10 (Inequalities with e).

$$e^x \geq 1+x \ \text{for } x \in \mathbb{R}$$
$$e^{-x} \leq 1 - \frac{x}{2} \ \text{for } 0 \leq x \leq 1$$
$$e^x \leq \frac{1}{1-x} \ \text{for } x < 1$$
$$\left(1 - \frac{1}{n}\right)^n \leq e^{-1} \leq \left(1 - \frac{1}{n}\right)^{n-1} \ \text{for } n \in \mathbb{N}$$

Proposition A.11 (Binomial coefficients). *Let $n \geq k \geq 0$. The binomial coefficients are defined as*

$$\binom{n}{k} = \binom{n}{n-k} = \frac{n!}{k!(n-k)!},$$

and it holds

$$\left(\frac{n}{k}\right)^k \leq \binom{n}{k} \leq \frac{n^k}{k!} \leq \left(\frac{ne}{k}\right)^k.$$

Proposition A.12 (Harmonic sum). *Let $H_n = \sum_{i=1}^n 1/i$ be the nth Harmonic sum. Then for any $n \in \mathbb{N}$*

$$H_n = \ln n + \Theta(1).$$

Proposition A.13 (Coupon collector's theorem). *In the coupon collector's problem, n different coupons are given and at each trial a coupon is chosen uniformly at random. Let X be a random variable describing the number of trials required to choose each coupon at least once. Then*

$$E(X) = nH_n$$

holds, where H_n denotes the nth Harmonic number, and

$$\lim_{n \to \infty} \text{Prob}(X \leq n(\ln n - c)) = e^{-e^c}$$

holds for each constant $c \in \mathbb{R}$.

References

Aarts E, Lenstra J K (2003) Local Search in Combinatorial Optimization. Wiley

Aleliunas R, Karp R M, Lipton R J, Lovász L, Rackoff C (1979) Random walks, universal traversal sequences, and the complexity of maze problems. In: Proceedings of the 20th Annual Symposium on Foundations of Computer Science (FOCS '79), IEEE Press, 218–223

Attiratanasunthron N, Fakcharoenphol J (2008) A running time analysis of an ant colony optimization algorithm for shortest paths in directed acyclic graphs. Information Processing Letters 105(3):88–92

Barahona F, Pulleyblank W (1987) Exact arborescences, matchings and cycles. Discrete Applied Mathematics 16:91–99

Bast H, Funke S, Sanders P, Schultes D (2007) Fast routing in road networks with transit nodes. Science 316(5824):566

Baswana S, Biswas S, Doerr B, Friedrich T, Kurur P P, Neumann F (2009) Computing single source shortest paths using single-objective fitness functions. In: Proceedings of the 10th International Workshop on Foundations of Genetic Algorithms (FOGA '09), ACM Press, 59–66

Beier R, Vöcking B (2004) Random knapsack in expected polynomial time. Journal of Computer and System Sciences 69(3):306–329

Briest P, Brockhoff D, Degener B, Englert M, Gunia C, Heering O, Jansen T, Leifhelm M, Plociennik K, Röglin H, Schweer A, Sudholt D, Tannenbaum S, Wegener I (2004) Experimental supplements to the theoretical analysis of eas on problems from combinatorial optimization. In: Proceedings of Parallel Problem Solving from Nature VIII (PPSN '04), volume 3242 of Lecture Notes in Computer Science, Springer, 21–30

Brockhoff D, Friedrich T, Neumann F (2008) Analyzing hypervolume indicator based algorithms. In: Proceedings of Parallel Problem Solving from Nature X (PPSN '08), volume 5199 of Lecture Notes in Computer Science, Springer, 651–660

Broder A (1989) Generating random spanning trees. In: Proceedings of the 30th Annual Symposium on Foundations of Computer Science (FOCS '89), IEEE Press, 442–447

Bui T N, Chaudhuri S, Leighton F T, Sipser M (1984) Graph bisection algorithms with good average case behavior. In: Proceedings of the 25th Annual Symposium on Foundations of Computer Science (FOCS '84), IEEE Press, 181–192

Chen J, Kanj I A, Xia G (2006) Improved parameterized upper bounds for vertex cover. In: Proceedings of the 31st International Symposium on Mathematical Foundations of Computer Science (MFCS '06), volume 4162 of Lecture Notes in Computer Science, Springer, 238–249

Chvátal V (1979) A greedy heuristic for the set-covering problem. Mathematics of Operations Research 4(3):233–235

Coello Coello C A, Van Veldhuizen D A, Lamont G B (2007) Evolutionary Algorithms for Solving Multi-Objective Problems. Springer, 2nd edition

Coffman E G Jr, Whitt W (1995) Recent asymptotic results in the probabilistic analysis of schedule makespans. In: Chrétienne P, Coffman E G, Lenstra J K, Liu Z(eds.) Scheduling Theory and its Applications, Wiley, 15–31

Colorni A, Dorigo M, Maniezzo V (1992) An investigation of some properties of an "ant algorithm". In: Proceedings of Parallel Problem Solving from Nature II (PPSN '92), Elsevier, 515–526

Cormen T, Leiserson C, Rivest R, Stein C (2001) Introduction to Algorithms. McGraw-Hill, 2nd edition

David H A, Nagaraja H N (2003) Order Statistics. Wiley, 3rd edition

Deb K (2001) Multi-objective optimization using evolutionary algorithms. Wiley

Doerr B, Happ E, Klein C (2007a) A tight analysis of the (1+1)-EA for the single source shortest path problem. In: Proceedings of the IEEE Congress on Evolutionary Computation (CEC '07), IEEE Press, 1890–1895

Doerr B, Happ E, Klein C (2008) Crossover can provably be useful in evolutionary computation. In: Proceedings of the Genetic and Evolutionary Computation Conference (GECCO '08), ACM Press, 539–546

Doerr B, Hebbinghaus N, Neumann F (2007b) Speeding up evolutionary algorithms through asymmetric mutation operators. Evolutionary Computation 15(4):401–410

Doerr B, Johannsen D (2007) Adjacency list matchings — an ideal genotype for cycle covers. In: Proceedings of the Genetic and Evolutionary Computation Conference (GECCO '07), ACM Press, 1203–1210

Doerr B, Neumann F, Sudholt D, Witt C (2007c) On the runtime analysis of the 1-ANT ACO algorithm. In: Proceedings of the Genetic and Evolutionary Computation Conference (GECCO '07), ACM Press, 33–40

Doerr B, Theile M (2009) Improved analysis methods for crossover-based algorithms. In: Proceedings of the Genetic and Evolutionary Computation Conference (GECCO '09), ACM Press, 247–254

Dorigo M, Blum C (2005) Ant colony optimization theory: A survey. Theoretical Computer Science 344:243–278

Dorigo M, Maniezzo V, Colorni A (1991) The ant system: An autocatalytic optimizing process. Technical Report 91-016 Revised, Politecnico di Milano

Dorigo M, Stützle T (2004) Ant Colony Optimization. MIT Press

Downey R G, Fellows M R (1999) Parameterized Complexity. Springer

Droste S, Jansen T, Wegener I (2002) On the analysis of the (1+1) evolutionary algorithm. Theoretical Computer Science 276:51–81

Duarte A, Sánchez Á, Fernández F, Cabido R (2005) A low-level hybridization between memetic algorithm and VNS for the max-cut problem. In: Proceedings of the Genetic and Evolutionary Computation Conference (GECCO '05), ACM Press, 999–1006

Edmonds J, Johnson E L (1973) Matching, Euler tours and the Chinese postman. Mathematical Programming 8:88–124

Ehrgott M (2005) Multicriteria optimization. Springer, 2nd edition

Eiben A, Smith J (2007) Introduction to Evolutionary Computing. Springer

El-Fallahi A, Prins C, Calvo R W (2008) A memetic algorithm and a tabu search for the multi-compartment vehicle routing problem. Computers and Operations Research 35(5):1725–1741

Euler L (1741) Solutio problematis ad geometriam situs pertinentis. Commentarii academiae scientiarum Petropolitanae 8:128–140

Farooq M (2008) Bee-Inspired Protocol Engineering. Springer

Feige U (1998) A threshold of $\ln n$ for approximating set cover. Journal of the ACM 45(4):634–652

Feller W (1968) An Introduction to Probability Theory and Its Applications, volume 1. Wiley, 3rd edition

Feller W (1971) An Introduction to Probability Theory and Its Applications, volume 2. Wiley, 2nd edition

Fogel L, Owens M, Walsh M (1966) Artificial Intelligence through simulated evolution. Wiley

Frenk J B G, Rinnooy Kan A H G (1986) The rate of convergence to optimality of the LPT rule. Discrete Applied Mathematics 14:187–197

Frenk J B G, Rinnooy Kan A H G (1987) The asymptotic optimality of the LPT rule. Mathematics of Operations Research 12(2):241–254

Friedrich T, He J, Hebbinghaus N, Neumann F, Witt C (2009) Analyses of simple hybrid evolutionary algorithms for the vertex cover problem. Evolutionary Computation 17(1):3–20

Friedrich T, Hebbinghaus N, Neumann F, He J, Witt C (2007) Approximating covering problems by randomized search heuristics using multi-objective models. In: Proceedings of the Genetic and Evolutionary Computation Conference (GECCO '07), ACM Press, 797–804, extended version to appear in Evolutionary Computation

Garnier J, Kallel L, Schoenauer M (1999) Rigorous hitting times for binary mutations. Evolutionary Computation 7(2):173–203

Giel O (2003) Expected runtimes of a simple multi-objective evolutionary algorithm. In: Proceedings of the IEEE Congress on Evolutionary Computation (CEC '03), IEEE Press, 1918–1925

Giel O, Wegener I (2003) Evolutionary algorithms and the maximum matching problem. In: Procedings of the 20th Annual Symposium on Theoretical Aspects of Computer Science (STACS '03), volume 2607 of Lecture Notes on Computer Science, Springer, 415–426

Giel O, Wegener I (2004) Searching randomly for maximum matchings. In: Electronic Colloquium on Computational Complexity (ECCC), report no. 76

Giel O, Wegener I (2006) Maximum cardinality matchings on trees by randomized local search. In: Proceedings of the Genetic and Evolutionary Computation Conference (GECCO '06), ACM Press, 539–546

Gottlieb J, Julstrom B A, Raidl G R, Rothlauf F (2001) Prüfer numbers: A poor representation of spanning trees for evolutionary search. In: Proceedings of the Genetic and Evolutionary Computation Conference (GECCO '01), Morgan Kaufmann, 343–350

Graham R L (1969) Bounds on multiprocessing timing anomalies. SIAM Journal on Applied Mathematics 17:263–269

Gutjahr W J (2007) Mathematical runtime analysis of ACO algorithms: Survey on an emerging issue. Swarm Intelligence 1:59–79

Hajek B (1982) Hitting-time and occupation-time bounds implied by drift analysis with applications. Advances in Applied Probability 13(3):502–525

Hierholzer C (1873) Über die Möglichkeit, einen Linienzug ohne Wiederholung und ohne Unterbrechung zu umfahren. Mathematische Annalen 6:30–32

Hochbaum D (1997) Appromixation Algorithms for NP-hard Problems. PWS Publishing Company

Holland J H (1975) Adaptation in Natural and Artificial Systems. University of Michigan Press

Hoos H H, Stützle T (2004) Stochastic Local Search: Foundations & Applications. Elsevier/Morgan Kaufmann

Hopcroft J E, Karp R M (1973) An $n^{5/2}$ algorithm for maximum matchings in bipartite graphs. SIAM Journal on Computing 2:225–231

Horoba C, Neumann F (2008) Benefits and drawbacks for the use of epsilon-dominance in evolutionary multi-objective optimization. In: Proceedings of the Genetic and Evolutionary Computation Conference (GECCO '08), ACM Press, 641–648

Horoba C, Neumann F (2009) Additive approximations of Pareto-optimal sets by evolutionary multi-objective algorithms. In: Proceedings of the 10th International Workshop on Foundations of Genetic Algorithms (FOGA '09), ACM Press, 79–86

Horoba C, Sudholt D (2009) Running time analysis of ACO systems for shortest path problems. In: Proceedings of the 2nd International Workshop on Engineering Stochastic Local Search Algorithms (SLS '09), volume 5752 of Lecture Notes in Computer Science, Springer, 76–91

Jansen T, Sudholt D (2005) Design and analysis of an asymmetric mutation operator. In: Proceedings of the IEEE Congress on Evolutionary Computation (CEC '05), IEEE Press, 497–504

Jansen T, Wegener I (2001) Evolutionary algorithms – how to cope with plateaus of constant fitness and when to reject strings of the same fitness. IEEE Transactions on Evolutionary Computation 5(6):589–599

Jerrum M, Sorkin G B (1998) The Metropolis algorithm for graph bisection. Discrete Applied Mathematics 82(1–3):155–175

Kano M (1987) Maximum and k-th maximal spanning trees of a weighted graph. Combinatorica 7(2):205–214

Karakostas G (2005) A better approximation ratio for the vertex cover problem. In: Proceedings of the 32nd International Colloquium on Automata, Languages and Programming (ICALP '05), volume 3580 of Lecture Notes in Computer Science, Springer, 1043–1050

Karger D R, Klein P N, Tarjan R E (1995) A randomized linear-time algorithm to find minimum spanning trees. Journal of the ACM 42(2):321–328

Kennedy J, Eberhart R C (1995) Particle Swarm Optimization. In: Proceedings of the IEEE International Conference on Neural Networks, volume 4, IEEE Press, 1942–1948

Kim S J, Choi M K (2007) Evolutionary algorithms for route selection and rate allocation in multirate multicast networks. Applied Intelligence 26(3):197–215

Knowles J D, Corne D (2001) A comparison of encodings and algorithms for multiobjective spanning tree problems. In: Proceedings of the IEEE Congress on Evolutionary Computation (CEC '01), IEEE Press, 544–551

Korte B, Vygen J (2005) Combinatorial Optimization: Theory and Algorithms. Springer, 3rd edition

Koza J R (1991) Evolving a computer program to generate random numbers using the genetic programming paradigm. In: Proceedings of the 4th International Conference on Genetic Algorithms (ICGA '91), Morgan Kaufmann, 37–44

Kratsch S, Neumann F (2009) Fixed-parameter evolutionary algorithms and the vertex cover problem. In: Proceedings of the Genetic and Evolutionary Computation Conference (GECCO '09), ACM Press, 293–300

Kruskal J B (1956) On the shortest spanning subtree of a graph and the traveling salesman problem. In: Proceedings of the American Mathematical Society, volume 7, 48–50

Lacomme P, Prins C, Ramdane-Chérif W (2001) A genetic algorithm for the capacitated arc routing problem and its extensions. In: Proceedings of Applications of Evolutionary Computing, EvoWorkshops 2001: EvoCOP, EvoFlight, EvoIASP, EvoLearn, and EvoSTIM, volume 2037 of Lecture Notes in Computer Science, Springer, 473–483

Laumanns M, Thiele L, Deb K, Zitzler E (2003) Combining convergence and diversity in evolutionary multiobjective optimization. Evolutionary Computation 10(3):263–282

Laumanns M, Thiele L, Zitzler E (2004) Running time analysis of multiobjective evolutionary algorithms on pseudo-boolean functions. IEEE Transactions on Evolutionary Computation 8(2):170–182

Liang K H, Yao X, Newton C S, Hoffman D (2002) A new evolutionary approach to cutting stock problems with and without contiguity. Computers and Operations Research 29(12):1641–1659

Mayr E W, Plaxton C G (1992) On the spanning trees of weighted graphs. Combinatorica 12(4):433–447

Mehlhorn K, Sanders P (2008) Algorithms and Data Structures: The Basic Toolbox. Springer

Micali S, Vazirani V V (1980) An $O(\sqrt{|V|} \cdot |E|)$ algorithm for finding maximum matching in general graphs. In: Proceedings of the 21st Annual Symposium on Foundations of Computer Science (FOCS '80), IEEE Press, 17–27

Michalewicz Z (1995) A survey of constraint handling techniques in evolutionary computation methods. In: Evolutionary Programming, 135–155

Michalewicz Z, Fogel D B (2004) How to solve it: Modern heuristics. Springer

Motwani R, Raghavan P (1995) Randomized Algorithms. Cambridge University Press

Mühlenbein H (1992) How genetic algorithms really work: Mutation and hillclimbing. In: Proceedings of Parallel Problem Solving from Nature II (PPSN '92), Elsevier, 15–26

Nemhauser G, Ullman Z (1969) Discrete dynamic programming and capital allocation. Management Science 15(9):494–505

Neumann F (2007) Expected runtimes of a simple evolutionary algorithm for the multi-objective minimum spanning tree problem. European Journal of Operational Research 181(3):1620–1629

Neumann F (2008) Expected runtimes of evolutionary algorithms for the Eulerian cycle problem. Computers and Operations Research 35(9):2750–2759

Neumann F, Reichel J (2008) Approximating minimum multicuts by evolutionary multi-objective algorithms. In: Proceedings of Parallel Problem Solving from Nature X (PPSN '08), volume 5199 of Lecture Notes in Computer Science, Springer, 72–81

Neumann F, Reichel J, Skutella M (2008) Computing minimum cuts by randomized search heuristics. In: Proceedings of the Genetic and Evolutionary Computation Conference (GECCO '08), ACM Press, 779–786

Neumann F, Sudholt D, Witt C (2009) Analysis of different MMAS ACO algorithms on unimodal functions and plateaus. Swarm Intelligence 3(1):35–68

Neumann F, Wegener I (2007) Randomized local search, evolutionary algorithms, and the minimum spanning tree problem. Theoretical Computer Science 378(1):32–40

Neumann F, Witt C (2009) Runtime analysis of a simple ant colony optimization algorithm. Algorithmica 54(2):243–255

Neumann F, Witt C (2010) Ant colony optimization and the minimum spanning tree problem. Theoretical Computer Science 411(25):2406–2413

Nocedal J, Wright S (2000) Numerical Optimization. Springer

Oliveto P S, He J, Yao X (2008) Analysis of population-based evolutionary algorithms for the vertex cover problem. In: Proceedings of the IEEE Congress on Evolutionary Computation (CEC '08), IEEE Press, 1563–1570

Oliveto P S, He J, Yao X (2009) Analysis of the (1+1)-EA for finding approximate solutions to vertex cover problems. IEEE Transactions on Evolutionary Computation 13(5):1006–1029

Oliveto P S, Witt C (2008) Simplified drift analysis for proving lower bounds in evolutionary computation. In: Proceedings of Parallel Problem Solving from Nature X (PPSN '08), volume 5199 of Lecture Notes in Computer Science, Springer, 82–91

Papadimitriou C H, Steiglitz K (1998) Combinatorial Optimization: Algorithms and Complexity. Dover

Papadimitriou C H, Yannakakis M (2000) On the approximability of trade-offs and optimal access of web sources. In: Proceedings of the 41st Annual Symposium on Foundations of Computer Science (FOCS '00), IEEE Press, 86–92

Prim R C (1957) Shortest connection networks and some generalizations. Bell System Technical Journal 36:1389–1401

Puchinger J, Raidl G R, Koller G (2004) Solving a real-world glass cutting problem. In: Proceedings of the 4th European Conference on Evolutionary Computation in Combinatorial Optimization (EvoCOP '04), volume 3004 of Lecture Notes in Computer Science, Springer, 165–176

Raidl G R, Julstrom B A (2003) Edge sets: an effective evolutionary coding of spanning trees. IEEE Transactions on Evolutionary Computation 7(3):225–239

Raz R, Safra S (1997) A sub-constant error-probability low-degree test, and a sub-constant error-probability PCP characterization of NP. In: Proceedings of the 29th Annual ACM Symposium on the Theory of Computing (STOC '97), ACM Press, 475–484

Rechenberg I (1973) Evolutionsstrategie – Optimierung technischer Systeme nach Prinzipien der biologischen Evolution. Frommann-Holzboog

Reichel J, Skutella M (2007) Evolutionary algorithms and matroid optimization problems. In: Proceedings of the Genetic and Evolutionary Computation Conference (GECCO '07), ACM Press, 947–954

Reichel J, Skutella M (2009) On the size of weights in randomized search heuristics. In: Proceedings of the 10th International Workshop on Foundations of Genetic Algorithms (FOGA '09), ACM Press, 21–28

Rizzoli A E, Montemanni R, Lucibello E, Gambardella L M (2007) Ant colony optimization for real-world vehicle routing problems. Swarm Intelligence 1(2):135–151

Sanders P, Schultes D (2006) Engineering highway hierarchies. In: Proceedings of the 14th Annual European Symposium on Algorithms (ESA '06), volume 4168 of Lecture Notes in Computer Science, Springer, 804–816

Sasakik G H, Hajek B (1988) The time complexity of maximum matching by simulated annealing. Journal of the ACM 35:387–403

Scharnow J, Tinnefeld K, Wegener I (2004) The analysis of evolutionary algorithms on sorting and shortest paths problems. Journal of Mathematical Modelling and Algorithms 3(4):349–366

Schwefel H P (1981) Numerical optimization for computer models. Wiley

Sorkin G B (1991) Efficient simulated annealing on fractal energy landscapes. Algorithmica 6(3):367–418

Stützle T, Hoos H H (2000) Max-min ant system. Future Generation Computer Systems 16(8):889–914

Sudholt D, Witt C (2010) Runtime analysis of a binary particle swarm optimizer. Theoretical Computer Science 411(21):2084–2100

Swinscow T D V, Campbell M J (2001) Statistics at square one. BMJ Publishing Group, 10th edition

van Laarhoven P, Aarts E (1997) Simulated Annealing: Theory and Applications. Springer

Vazirani V (2001) Appromixation Algorithms. Springer

Wegener I (2005a) Complexity Theory – Exploring the Limits of Efficient Algorithms. Springer

Wegener I (2005b) Simulated annealing beats Metropolis in combinatorial optimization. In: Proceedings of the 32nd International Colloquium on Automata, Languages and Programming (ICALP '05), volume 3580 of Lecture Notes on Computer Science, Springer, 589–601

Wilson D B (1996) Generating random spanning trees more quickly than the cover time. In: Proceedings of the 28th Annual ACM Symposium on the Theory of Computing (STOC '96), ACM Press, 296–303

Witt C (2005) Worst-case and average-case approximations by simple randomized search heuristics. In: Proceedings of the 22th Symposium on Theoretical Aspects of Computer Science Proceedings (STACS '05), volume 3404 of Lecture Notes on Computer Science, Springer, 44–56

Witt C (2009) Greedy local search and vertex cover in sparse random graphs. In: Proceedings of the 6th Annual Conference on Theory and Applications of Models of Computation (TAMC '09), volume 5532 of Lecture Notes in Computer Science, Springer, 410–419

Wolpert D, Macready W G (1997) No free lunch theorems for optimization. IEEE Transactions on Evolutionary Computation 1(1):67–82

Zhou G, Gen M (1999) Genetic algorithm approach on multi-criteria minimum spanning tree problem. European Journal of Operational Research 114:141–152